DARWIN'S LOVE OF LIFE

DARWIN'S LOVE OF LIFE

A Singular Case of Biophilia

KAY HAREL

Columbia University Press
New York

Columbia University Press
Publishers Since 1893
New York Chichester, West Sussex
cup.columbia.edu

Paperback edition, 2024

Library of Congress Cataloging-in-Publication Data

Names: Harel, Kay, author.
Title: Darwin's love of life : a singular case of biophilia / Kay Harel.
Description: New York : Columbia University Press, [2022] |
Includes bibliographical references and index.
Identifiers: LCCN 2022015198 (print) | LCCN 2022015199 (ebook) |
ISBN 9780231208086 (hardback) | ISBN 9780231216708 (pbk.) |
ISBN 9780231557269 (ebook)
Subjects: LCSH: Darwin, Charles, 1809–1882. | Naturalists—Great
Britain—Biography. | Biology—Philosophy.
Classification: LCC QH31.D2 H37 2022 (print) | LCC QH31.D2
(ebook) | DDC 576.8/2—dc23/eng/20220511
LC record available at https://lccn.loc.gov/2022015198
LC ebook record available at https://lccn.loc.gov/2022015199

Cover design: Henry Sene Yee
Cover images: iStock.com

With gratitude to CR, life, love, and encouragement

Meanwhile, spring had come. . . . One of those rare springs that bring joy to plants, animals, and people alike.

—Leo Tolstoy

CONTENTS

PREFACE

FIGURE 0.1 Cherishing a rare flower.

Darwin posed for this pastel drawing with his older sister Caroline in 1816 at age six or seven. Decades later, after he overturned our understanding of life, Darwin's son Francis, who became a botanist, reminisced: "I used to like to hear him admire the beauty of a flower: it was a kind of gratitude to the flower itself, and a personal love for its delicate form and color. I seem to remember him gently touching a flower he delighted in; it was the same simple admiration that a child might have." Darwin shared his love for flowers with his wife, Emma, who also had quite the green thumb, and their home was a haven for flowers, as well as a veritable botanical research station. Orchids were Darwin's favorite. He devoted a tome to them, full of detail and passion: "The contrivances by which Orchids are fertilised . . . are almost as perfect as any of the most beautiful adaptations in the animal kingdom. . . . Orchids . . . rank amongst the most singular and most modified forms." The "entangled bank" Darwin depicted in the much-quoted last paragraph of *On the Origin of Species* was a real one, near his home, full of wild orchids, inspiring hours of scientific scrutiny and, always, evoking in him biophilia's simple admiration. In this drawing by Darwin family friend Ellen Sharples, Darwin cradles an exotic treasure, *Lachenalia aloides*, a native of southernmost Africa.

Source: Reproduced with permission from John van Wyhe, ed., *The Complete Work of Charles Darwin Online*, 2002–, http://darwin-online.org.uk/. The original is held in the Down House collection of the historical preservation group English Heritage.

MY BOOK IS a project of consilience, defined as "jumping together"—in my case, the joining of different forms of knowledge to gain new insights. In his 1998 book *Consilience: The Unity of Knowledge*, Edward O. Wilson wrote that "consilience is not yet science. It is a metaphysical view. . . . The strongest appeal of consilience is in the prospect of intellectual adventure and, given even modest success, the value of understanding the human condition with a higher degree of certainty." This book is, I hope, such an intellectual adventure. I draw on ideas from aesthetics, anthropology, astronomy, biology, evolutionary theories of many stripes, philosophy, physics, poetry, primatology, psychoanalysis, and theoretical physics. One line of poetry, one theory by a philosopher about the origins of the universe, have, for me, as much gravitas as a rigorous biological study or aesthetic analysis in helping to understand biophilia and its place in Darwin's life and work.

Biophilia—a love of life—is an obscure term of art and a vague concept, but it has a provenance in the thought of a few influential intellects, such as the psychoanalyst Erich Fromm and the modern naturalist Edward O. Wilson. Fromm based his idea of biophilia on the drive for survival, and he emphasized a love for whatever serves survival, including altruism. Wilson uses the term to describe the love of nature and the unity of all living organisms. Other theorists have proposed similar ideas in other terms. Biophilia is an "omnibus" term, as the philosopher Akeel Bilgrami points out, especially as figured in my study, where it includes everything from the reflexive survival instinct to the deliberate cultivation of beauty. There is no one-size-fits-all definition of biophilia, but it can be seen in characteristic patterns. There is an old saw in philosophy that what cannot be defined may still be fleshed out as a reality via description. That is my project here.

My guiding premise is that Darwin's evolutionary views, as described in his writings and analyzed in his private notebooks, are shaped by biophilia and that much of what he saw was biophilia in action. His singular case of biophilia primed him to see what others did not. Darwin did not use the word *biophilia*, but other words harking to the concept: "there is one instinct to all animals," "one thinking principle," "one living spirit." The Darwin scholar Howard E. Gruber wrote, "We need to understand what forms of thought he used." This book shows biophilia as one of those forms.

My book is a thought experiment, an imaginative exercise, an impressionistic painting in words about the perception and psyche of a giant in history. This is not an academic study, not an act of scholarship, but a personal book. Darwin praises the scientist Alexander von Humboldt for his "rare union of poetry with science." That is my goal and method in this book. I am neither fish nor fowl but a quirky hybrid chasing consilience. I fell in love with Darwin the minute I began reading *On the Origin of Species*. Between the lines, I saw someone who seemed extraordinary: careful with fact but sweeping with theory, charming but also logical, precise with language but in an embrace with the unknown, with a gentle voice but a bulldozer's forward momentum. My ever-present question when reading Darwin was: Who *is* this guy? This project was in service to a mystery in my mind. My ever-present question while writing this book was: Would Darwin think this is true?

The seven chapters constitute a set of interrelated essays on the theme of Darwin's biophilia. In "A Study in Biophilia," the first chapter, I introduce the concept and give an outline of Darwin's singular case of biophilia. Each of the next six chapters examines one facet of Darwin's work or life and how it was sculpted by his love of life. The topics are: dogs (chap. 2), facts (chap. 3), thought (chap. 4), emotion (chap. 5), and beauty (chap. 6). The final chapter (chap. 7) portrays Darwin's marriage and domestic world, focusing on his wife, the heiress Emma Wedgwood, as a woman who shared his biophilia. I peer into their partnership and home, which was full of a love of life that left its own legacy.

The topics are increasingly complex, and for the reader progressing straight through, they accrue to give a deepening understanding of how biophilia influences Darwin's work and life. The concept of biophilia likewise unfolds more and more over the course of the book. But other than the essential introductory chapter, the book is modular: readers can jump to whatever topic interests them without any loss of continuity.

In *The Structure of Scientific Revolutions*, Thomas S. Kuhn wrote, "Philosophers of science have repeatedly demonstrated that more than one theoretical construction can always be placed upon a given collection of data." After the placement, of course, scientists next examine the fit. I have placed my theory about biophilia over a collection of data. Please examine the fit.

DARWIN'S LOVE OF LIFE

Chapter One

A STUDY IN BIOPHILIA

We cannot divide man sharply into
an animal and a rational part.

—William James

FIGURE 1.1 "Ancient sage."

Darwin, age sixty-five, in an 1874 photograph by his son Leonard. Some two years later, Darwin sketched his autobiography as a private document intended for his children and grandchildren, confessing, for example, that he "was in many ways a naughty boy." More seriously, he wrote: "I have attempted to write . . . as if I were a dead man in another world looking back at my own life. Nor have I found this difficult, for life is nearly over with me." (He died six years later.) In biophilia, a profound love of life encompasses the pragmatic acceptance of death. A few months after Darwin drafted his memoir, the Russian botanist Kliment Timiryazev visited the Darwin home and later compared Darwin to an "ancient sage or Old Testament patriarch." Describing his studies of chlorophyll, Timiryazev was thrilled when Darwin replied, "Chlorophyll is perhaps the most interesting organic substance." Indeed, converting the physical energy of light into the biochemical energy of a plant, chlorophyll shows that the inorganic can become organic, a most interesting point indeed for sages and others who contemplate life's origin.

Source: Reproduced with permission from John van Wyhe, ed., *The Complete Work of Charles Darwin Online*, 2002–, http://darwin-online.org.uk/.

A BROKEN CLOCK is right twice a day. So it happens that those crackpot phrenologists identified a large "organ of veneration" on Charles Darwin's head. But Darwin's reverence was not for the Judeo-Christian god worshipped by his wife, Emma. Indeed, their religious disparity strained the Darwins' otherwise robust marriage. Charles and Emma realized early on that Darwin was no churchgoing soul. In his memoir for his children, he commented, "Considering how fiercely I have been attacked by the orthodox, it seems ludicrous that I once intended to be a clergyman." Ironic maybe, but not so ludicrous given the size of his bump of reverence. It is possible that the bump grew during Darwin's trip around the world on the HMS *Beagle*, begun at age twenty-two. On Darwin's return, his father exclaimed, "Why the shape of his head is quite altered!" as Darwin recounted their reunion in the memoir. It is true that he had just observed five years' worth of mind-bending sights, a world's worth of variety and uniformity, growth and death, horror and beauty, mystery upon mystery, all stowed in his protean brain.

What Darwin saw put him in mind of "one stock," "all . . . obeying one law." He mused, "There is one living spirit . . . which assumes a multitude of forms." So he scrawled in his earliest notebooks, stammering intuitions in private. "One living spirit" sounds like a god, speaking of reverence, perhaps of the kind divined by Indigenous peoples since religion evolved. Or perhaps Darwin retained the phrase from the poet Percy Bysshe Shelley's *Hellas: A Lyrical Drama*. In Darwin's study in a folder labeled "Old and Useless Notes" was this scrap: "There is one instinct to all animals modified according to species."

That statement has the ring of a proclamation, even if he did not follow up. His other ideas about that "one stock" had his attention.

The one instinct Darwin sensed was, perhaps, biophilia, a drive for life that encompasses everything from the reflex to survive to the desire for beauty, "modified according to species." Biophilia—a Latinate term for "love of life"—as a natural force is not proven to exist, and it is a nebulous theory. But self-preservation is the first law of nature, as we were told centuries ago by the novelist and evolutionary theorist Samuel Butler. Self-preservation is biophilia writ plain and biological. Most recently, the Harvard naturalist Edward O. Wilson has profiled human biophilia as an affinity for nature, which is a facet of the love of life. Other sketches of biophilia have been drawn, and other theories posit similar forces. Darwin's one living spirit, one instinct, one law, accords with the Gestalt of biophilia.

Darwin was primed to note the love of life because he himself had a singular case of biophilia. William James remarked: "What is shown to be true in a marked degree of some persons is probably true in some degree of all, and may in a few be true in an extraordinarily high degree." Darwin was naturally endowed with an intuitive sense of life's tautological pursuit of life, whether he was channeling the labors of the earthworm, the mating battles of male salmon, the singing of church choirs, the stories of Mark Twain, a heliotropic plant twisting toward the sun, a baboon rescuing a relative, or a young man dancing at a country fair. He saw his one living spirit on every tangled bank he contemplated, in the character of dogs and the glory of facts, and throughout the varieties of passionate and aesthetic experience. His biophilia was the lens always before his eyes as he examined every nook and cranny of nature, culture, morality, beauty, anything, and everything. And he had a reverence for life's love of life, for the lifing of life.

Scholars have described facets of Darwin's biophilia, but not as such. He has many traits that mark biophilia: originality, intuition, insight, empathy with many others, creative thought, wisdom, a sense of oneness, the ability to envision truths without proofs—all these result from a profound sympathy with the impulse and logic of life. Howard E. Gruber wrote that Darwin had "taken the whole province of living things as his subject matter." Ernst Mayr lauded his "exceptionally fertile mind," able to grow theoretical trees from real seeds. Bernard Campbell lauded his "surprisingly accurate conclusions." Indeed, an entire parlor game could be designed called "Darwin Guessed Right." Gavin de Beer wrote of his "uncanny dumb sagacity." C. M. Yonge: "An imagination that ranged freely through space and time." George Levine: "In Darwin, intellect and feeling were one." Janet Browne: "He stood each problem on its head until it was not so much of a problem at all." Adrian Desmond and James Moore pointed out his "brazen speculation." Even Darwin's contemporary Saint George Jackson Mivart, who trashed *The Descent of Man*, praised his "sympathetically understanding observation." And as Gruber declared, he could "consider so deeply and so unflinchingly the whole range of possibilities." These encomiums point at his biophilia.

Gruber asserted that "we need to understand what forms of thought he used." Darwin's biophilia is one such form, a distinctive cast of thought, the slant of the truth in his scientific soul.

He saw evolutionary truths self-evident in the light of biophilia but obscure to the theories prevalent in his day. His biophilia explains why he did not distinguish between "high" and "low" life forms, the continuity he traced between plant and mathematician, and the way he saw reason as the descendent of instinct. He saw the logic of biophilia playing out everywhere.

INTRODUCING BIOPHILIA

Biophilia is no crackpot hypothesis, no cousin of phrenology. Its lack of proof is temporary. That can be the case for the best of theories, as Darwin's own case proves. Biophilia is on the compass of the scientist, the theoretician, the psychologist, and the odd Icelandic singer—the avant-garde icon Björk created a multimedia project entitled *Biophilia* that won a Grammy award in 2013. Biophilia has appeal, gravitas, and its own coherence.

If self-preservation is the first law of nature, as Butler declared, it is also the first act of biophilia. Put another way: biophilia originates as life's drive to live. To borrow from the poet Kahlil Gibran, it may be understood most simply as "life's longing for itself." Darwin was often drawn to manifestations of the survival instinct, the pursuit of living. One day during his trip around the world, he was mesmerized by a "small and pretty kind of spider," he wrote, who "feigned death" when frightened. Darwin was awed by the strategy, the complex and ready reflex. Biophilia describes the spider who does not *decide* to defend its life but just *does*. We can not just decide to stop breathing; when suffocating, our bodies struggle for air. The need to breathe is connatural. The body strives to live. This drive to live—the visceral love of life—accords with his "one instinct."

The psychoanalyst Erich Fromm rooted his understanding of biophilia on self-preservation. He stood on the shoulders of "biologists and philosophers," Fromm wrote, as he considered the "inherent quality of all living substance to live, to preserve its existence." Fromm's theory figures biophilia as a psychological love of life at one end of a continuum at the other end of which is a psychological love of death:

> The full unfolding of biophilia is to be found in the productive orientation. The person who fully loves life is attracted by the process of life and growth in all its spheres. He prefers to construct rather than to retain. He is capable of wondering, and he prefers to see something new to the security of finding confirmation of the old. He loves the adventure of living more than he does certainty. . . . He sees the whole rather than only the parts. . . . He wants to mold and to influence by love, reason, by his example. . . . Biophilic ethics have their own principle of good and evil. Good is all that serves life, evil is all that serves death. Good is reverence for life, all that enhances life, growth, unfolding. Evil is all that stifles life, narrows it down, cuts it to pieces. Joy is virtuous and sadness is sinful. . . . The pure necrophile is insane; the pure biophile is saintly.

Darwin's biophilia aligns with Fromm's definition from the first sentence, with the adjective "productive." Darwin's output was prodigious; he wrote dozens of volumes of experimental, taxonomic, and theoretical contributions, on a range of topics from barnacles and orchids to earthworms and emotions to his most famous, evolution. Fromm next specified that the biophile is "attracted to the processes of life and growth in all its spheres." Darwin chose to study nature. His attraction to the processes of life was like a gravitational pull. From his childhood in rural England, he lived and breathed kinship with all of nature. Starting with collections of rocks and beetles, he became a lifelong connoisseur of nature's artifacts, processes, and kaleidoscope of life's forms. He elucidated "growth in all . . . spheres," from plants to adaptations. Darwin rejected other educational tracks and career opportunities as he drew toward the study of life and the nascent profession of science. He spent decades in his study and homemade laboratories, where he studied countless thousands of specimens and fashioned fact after hard-won fact.

Synchronizing with Fromm's next point that the biophile "prefers to construct": for all Darwin's bench skills, he preferred theorizing, once confessing that he loved hypothesis: "I can not resist forming one on every subject." He privately referred to his penchant for theorizing as "building castles in the air." He was serious about this daydreaming: "Such trains of thought make a discoverer," he believed. He was inclined to envision the whole of evolution in the synecdoche of a bird's beak. In this, he continues to accord with Fromm's definition of a biophile as a person who "sees the new and the whole rather than the parts." Darwin synthesized quite a few wholes. He said that *On the Origin of Species* was "one long argument." His theory of evolution was a huge whole that required four theoretically contiguous volumes: *On the Origin of Species by*

Means of Natural Selection, *The Variation of Animals and Plants Under Domestication*, *The Descent of Man and Selection in Relation to Sex*, and *The Expression of the Emotions in Man and Animals*. Each is a doorstopper.

Fromm also characterized the biophile as a person who wonders and discovers. Walking through a forest in Brazil evoked adoration in Darwin, who wrote of "higher feelings of wonder, astonishment, and devotion, which fill and elevate the mind." His son Francis reminisced about seeing his father so transfixed: "I used to like to hear him admire the beauty of a flower: it was a kind of gratitude to the flower itself, and a personal love for its delicate form and color. I seem to remember him gently touching a flower he delighted in; it was the same simple admiration that a child might have." Wonder was his default state. He articulated his stance in *On the Origin of Species*: "Beings now ranked as very low in the scale . . . [have] really wondrous and beautiful organisation." And as for the discovering, most of the world today recognizes that Darwin worked far ahead of the frontiers of knowledge.

Darwin was ever able to, as Fromm wrote, "see something new." It is axiomatic that only the curious see the new, and Darwin's curiosity was extravagant. He was a lackadaisical student at Cambridge, but a professor nevertheless remarked, "What a fellow that Darwin is for asking questions." All his life, he relished the unknown, marveled at the ordinary and the bizarre, and was lured by mysteries such as yawns and peacock feathers. The Darwin scholar Gruber noted that he felt "the quest for truth as a source of pleasure." He loved learning.

Another intellectual component of biophilia that Fromm highlighted is the use of reason and example. This is clear in Darwin's prose: "But how, it may be asked, can any analogous principle apply in nature? I believe it can and does apply . . . though it was a long time before I saw how." Confessing puzzlement followed by comprehension, he optimistically leads by example. And here is Darwin appealing to reason in *Origin*: "If man can by patience select variations useful to him, why . . . should not variations useful to nature's living products often arise, and be preserved or selected? What limit can be put to this power . . . ?" He plies the rhetoric "to mold and to influence by . . . his example," as Fromm wrote.

As for the more emotional aspects of biophilia, Darwin seems at one with Fromm's concept that "joy is virtuous" when he mulled over "virtuous happiness." Darwin's sense of the sanctity of happiness was clear in his love for one of his daughters, Annie, a particularly playful and kind girl who died young (see chapter 5). And in regard to his love for "the adventure of living more than . . . certainty," in Fromm's phrase: theorizing is, of course, an intellectual adventure. And Darwin's physical zest for adventure led him to voyage around the world.

Fromm asserts too that in biophilia, "good is reverence for life." Elements of this reverence—a deep respect for life, admiration, love, a worship of nature's magic—weave in and out of Darwin's writing. Of "Nature," he kowtowed, "We are too blind to understand her meaning," a statement reminiscent of what some religions assert about their gods. "I can see no limit to the amount of change . . . which may have been effected . . .

through nature's power," he wrote, again showing faith in an infinity that most associate with a reigning supernatural being. Reverence is a constant in Darwin when it comes to what he calls the "beauty and infinite complexity" of the adaptations of life forms. Time and again, he sounded awe: no one, he declared, "can study any living creature, however humble, without being struck with enthusiasm at its marvellous structure and properties." When addressing the complexity of the eye and its evolutionary antecedents, he wrote: "Wonderful changes in function are . . . possible."

Finally, the last sentence of Fromm's definition brings in the word "saintly." In his home and scientific conduct, Darwin was a sort of saint, calm and fair as he puttered around his house, pursuing truth and joy, indulging his children, and cherishing his wife. He obliged his neighbors, corresponded generously with scientists, dealt patiently with critics, and behaved like a gentleman even as he shattered conventional wisdom.

An appealing portrait of a benignant Darwin was written by Kliment Timiryazev, a brilliant botanist and plant physiologist who finagled his way to Darwin's estate in 1877. He recounted the story in a Russian journal in 1909 (and today we may stumble across it here and there on the internet). Timiryazev had procured a letter of introduction to Francis, who was by then a famous botanist, living and conducting research in the family home. Despite the letter, Timiryazev wrote that he still expected he would "see a door slammed into my face," as Darwin was by then an invalid. But toting an autographed copy of his own Russian book on Darwin, meant as a gift for Darwin, Timiryazev made the journey, a "shabby foreigner," he described himself.

He was, in fact, a cultural figure and a mover and a shaker: a Russian aristocrat who translated Charles Dickens and George Eliot, brought greenhouses to his country, and initiated higher-education programs for women. An asteroid and a crater on the far side of the moon are named for him.

Francis welcomed him, and, as luck favors the intrepid, Darwin wandered into the drawing room during the visit. Darwin postponed his nap for hours, his guest wrote, to discuss science over coffee and tea. Timiryazev wrote, "I couldn't help comparing him to an ancient sage or an Old Testament patriarch. . . . After our conversation began I saw him as a very kindly and gentle man and felt that I had known him for a long time. . . . The greatest scientist had turned out to be the most affable of men." For the Russian scientist, his host was the "humble organism" Darwin had lionized in *Origin* as having "marvellous . . . properties."

CONSANGUINE THEORIES

A scientific variety of human biophilia is delineated by the controversial naturalist Edward Wilson. He used the term to define his own love of the natural world—which led to accomplishments in multiple fields—as a deep urge "to explore and affiliate with life." Such an urge is central in Darwin's psyche. Biophilia and the natural sciences have synergy, of course. In characterizing the Nobel laureate Barbara McClintock's famous "feeling for the organism," the scholar Evelyn Fox Keller wrote, "Good science cannot proceed without a deep emotional investment on the part of the scientist. . . . McClintock's feeling for the organism is . . . a longing to embrace the world in its very being, through reason and beyond." That longing is biophilia. Fox Keller wrote that McClintock "always had an 'exceedingly strong feeling' for the oneness of things . . . a deep reverence for nature." McClintock famously said, "Everything is one," a rare moment of convergence for evolution, Buddhism, and biophilia. McClintock shares Darwin's prescience, oneness, and reverence; she was clearly a kindred spirit.

The "fundamental desire" described by the nineteenth-century poet Matthew Arnold also resonates with Fromm's biophilia. In his classic essay on "Literature and Science," he wrote of "the desire in men that good should for ever be present to them." This "fundamental desire" is at the root of "every impulse in us. . . . The instinct exists. Such is human nature. And . . . human nature is preserved by our following the lead of its innocent instincts. Therefore, in seeking to gratify this instinct in question, we are following the instinct of self-preservation in humanity," Arnold wrote. The basic elements of biophilia appear in this passage: attraction to the good, instinct, and survival.

Also outlining a form of biophilia is a biological theory that posits an intrinsic drive for life: the "living machines" or "autopoetic machines" delineated by the modern-day biologists Humberto Maturana and Francisco Varela. Autopoiesis means "self-creating." In their theory, living machines generate their futures—a definition of life itself. Darwin's "one thinking . . . principle" prefigures the autopoietic dictum that "living systems are cognitive systems, and living as a process is a process of cognition," a process of recognizing and responding to realities in the world—and this applies to amoebas and plants just as to *Homo sapiens sapiens.* That thought of some kind might be at the heart of the drive to survive is a theory with a tradition in evolutionary work. The modern scholar Robert J. Richards characterizes this tradition: "Early evolutionists . . . all proposed . . . that behavior and mind drove the evolutionary process. . . . They philosophically dissected nature and found mind at its core." And the subject is very much alive today, as described by Charles Lumsden and Wilson: "Cognition has become a burning issue even within the ranks of the theoretical physicists, a few of whom are convinced that the mind of the observer is somehow woven into the laws of subatomic particles." This is true on the simplest level imagined by William James when he points out that ideas are real and material from the moment they rearrange neural pathways in the brain. Mind is at the heart of matter in the argument of the modern evolutionary theorist Ernst Mayr when he writes, "Most acquisition of new structures in the course of evolution can be ascribed to selection forces exerted by newly acquired behaviors." All these theories—from the mind evolved by the mind to the autopoietic machine to the longing for life—are one biophilia.

For Darwin, pretty spiders, orchids, insectivorous plants, earthworms, and barnacles are cognizant beings, autopoietic. Their brains register light, temperature, food, and danger. Then in virtually mechanical sequence, these organic beings respond to their cognizance so as to live. They make choices—a "thinking principle," as Darwin said. For Darwin, "the most humble organism is much higher than the inorganic dust under our feet." Why? Because it seeks life. It has biophilia.

WHAT'S PHYSICS GOT TO DO WITH IT?

And the origin of biophilia, the turtle under the turtle? One theory offers a continuity from inorganic matter to the drive for life. According to the ideas of Charles Sanders Peirce, an American polymath of the nineteenth century, biophilia would have been humbly born in physics. Revered today, Peirce was a misfit who loitered in the halls of universities such as Harvard and Johns Hopkins, innovating in several disciplines by extending Darwin's evolutionary theory. The modern philosopher Daniel Dennett harks to such work when he writes that "the idea of evolution by natural selection unified the realm of life, meaning, and purpose with the realm of space and time, cause and effect, mechanism and physical law." Called a "thinker of eminent originality" by that thinker of eminent originality William James, Peirce spun seminal theories about the causes and workings of the world and the mind.

Peirce often thought in what he called the trichotomic, "the art of making three-fold divisions." This way of analysis begins with a "First," or "freshness," by which he means what is "original, spontaneous." A Firstness is a given—for example, in the present universe, gravity would be a Firstness. A reflexive imperative for survival is also a Firstness in all life forms—organic beings, as Darwin would say. Biophilia, like gravity, is a raw and lawful energy that just is. They are Firsts also in that they are "initiative" and "agent," and, indeed, both are prerequisite for life as we know it. If gravity initiates the fall of the apple, biophilia initiates our hunger for it.

Yet while gravity and biophilia may be Firstnesses on our planet, fundamental energies here and now, once upon a time they did not exist, on Peirce's view. They too have an evolutionary pedigree: "The only possible way of accounting for the laws of nature and for uniformity in general is to suppose them the results of evolution," he wrote. He believed that gravity had evolved in an earlier version of the cosmos, then governed the evolution of life. In that same evolutionary path, life emerged with its tautological will to live, its biophilia. In accord with Peirce's views, then, biophilia would have evolved from gravity, another law of life emerging from the laws of our universe. And there is a family resemblance between biophilia and gravity: Biophilia is life's attraction to life just as gravity is matter's attraction to matter. Biophilia replicates gravity. Biology recapitulates physics. This accords with Peirce's postulate that Nature repeats itself, using its best mechanisms—such as natural selection—time and again.

According to Peirce, the mind mimics the forces that made it: "Our minds having been formed under the influence of phenomena governed by the laws of mechanics, certain concepts entering into those laws become implanted in our minds." Echoing Peirce, the modern cosmologist John Barrow has written that the human mind results from "an evolutionary process that rewards faithful representations of reality with survival." Barrow elaborates: "Many of our attitudes toward the Universe and its contents, together with some of our own creations and fascinations, are subtle consequences of the structure of the Universe." Just so, what is given in the atom may be taken up by the cell, then given to the body, and eventually taken up by the brain—an organ renowned for imitation. And accordingly, the body's need for life might diversify and evolve into the mind's love of searching for and creating new and other forms of life—knowledge, beauty, joy, children—just as in the biological realm, "one species has given birth to clear and distinct species," as Darwin wrote in *On the Origin of Species.* The botanist Timiryazev made this very point in thinking about Darwin when he recounted his visit: "Darwin constantly, and more particularly in later years was leaning toward a new area of science, which if not a necessary component of 'Darwinism,' is its natural extension, as I have pointed out more than once." Curiosity is the natural child of the drive to survive.

Peirce's First encompasses such a diversification of biophilia: the First often has what he calls "virtual variety"; it is "full of life and variety." As Darwin noted regarding variety, in another phrase perfectly describing biophilia: "There is one thinking, sensible . . . principle . . . which is modified into endless forms, bearing a close relation in degree & kind to the endless forms of the living beings." Whereas in *Homo sapiens* a physical imperative and an impulse of the psyche may seem quite different, for Darwin, the difference is more like that between acorn and oak, fetus and adult—time changes form and expression. But they harbor one life force.

The more complex the species, the more complex the biophilia, and it is "modified according to species," to borrow

Darwin's words. Humans contain varieties of biophilia from the survival instinct on up, from the physical urgencies of air, food, and water to our psyche—with all our reverence for life, need for human connection, curiosity, love of truth, and attraction to beauty and to whatever enhances life. Just as the plant cannot choose to do osmosis, so too the baby does not choose to cry if hungry, the child cannot choose to enjoy laughing, the artist cannot choose otherwise than to create, and the scientist is ever curious. From so simple an impulse, "endless forms . . . evolved," to quote Darwin.

So this is the world according to Peirce: physics is seminal for life; nature repeats itself; the logic of the life force is shaped by physics; evolution shapes the mind, which works just as physics does; and so we have an ultraevolved mind mimicking the physics of our universe. Peirce's point of view is in accord with that of Wilson, who argued in his 1998 book *Consilience: The Unity of Knowledge* that "all tangible phenomena, from the birth of the stars to the workings of social institutions, are based on material processes that are ultimately reducible, however long and tortuous the sequences, to the laws of physics." This is, he wrote, the "central idea of consilience," which is a practice of combining the knowledge and insights of diverse disciplines and arts to create a synergetic analysis that increases human understanding. The present portrait of Darwin's biophilia is a project in consilience, then, by being eclectic and interdisciplinary and by resting on physics.

One of the basic facts of life for the universe is that it is filled with dualities and contradictions. The cosmologist John Barrow pointed out a basic contradiction: "The world is full of complex structures and erratic events that are the outcomes of a small number of simple and symmetrical laws. This is how the Universe can be, at once, simple and complicated." On this view, simplicity gives rise to its opposite, complexity. As in physics, so in evolution: initial parameters are few, but the laws of nature and the events of circumstance combine to produce countless results. Water, salt, and heat become, with time, the residents of New York City. Evolution generates the simple and the complex, the nearly eternal cockroach and the barren seed conjured by the genetic engineer. Biology is full of contradictions.

Darwin perceived that evolutionary processes can be contradictory: an adapted population of a particular living form changes its environment over time, and vice versa. He never asked which process is true; "both happen" was an answer for him. Others argued. But he saw that chance can create order, a paradox that legions could not and will not accept. He understood that in evolutionary stories every effect becomes in due time a cause—among humans, this transformation has spiraled; as the modern theorists Lumsden and Wilson note, "Culture is created and shaped by biological processes while the biological processes are simultaneously altered in response to cultural change." So it is that peculiar nonbinaries extend to *Homo sapiens*.

At the same time that Darwin perceived novel dualities and contradictions, he likewise perceived false dualities. His biophilia always sensed truth. He saw no duality between mind and body but saw that mind grows from body and may in time turn and reshape it. He did not place in opposition "rational" and "emotional" but saw that the rational may grow from the emotional, an idea taken up by William James. He did not view as normative the two-sex, male-dominant sociosexual system of *Homo sapiens* but knew it was but one variation on biology's rainbow theme. He did not see as opposites the need to survive and the sense of beauty.

Many people can tolerate neither bothness nor paradox. And scientists generally aim at straightforward cause and effect. But Darwin had the ability to register paradox, as when he speculated that "thought originated in sensation," told us that it is "always advisable to perceive clearly our own ignorance," and railed against "false facts." In his biophilia, Darwin was on good terms with the kind of oxymoron that might drive others mad or at least provoke charges of incoherence or implausibility.

The interdisciplinary thinker Alexander Argyros reminds us that "the engine for cosmic evolution is the tendency of the universe to get itself tangled up in unresolvable paradoxes." This corroborates the view of biophilia as a child of physics that encompasses the paradox of self-contradictory duality. Darwin saw all the paradox in the coexistence of novel *and* typical, chance *and* pattern. He saw product and process as circular. He saw that Nature, including *Homo sapiens*, copies *and* creates. So it is no surprise that Darwin posited two contradictory forces propelling evolution—uniformity and variety. Survival happens and species evolve depending on uniformity—in signals, surroundings, food sources, and, as we know today, genes—and on variety—in mutations, adaptation, and circumstances. Evolution is at once predictable yet generates infinite variation. At work are both laws and "flaws," Barrow quipped, writing of the evolution of "a Universe governed by a small number of symmetrical laws [able] to manifest an infinite diversity of complex, asymmetrical states." We live, he wrote, in a "cosmic environment within which the logic of natural selection has allowed the hand of time to fashion living complexity." And when Barrow wrote that natural selection created our "cosmic environment," he builds on Peirce's assertion that natural selection was at play, at work, in the universe.

Spend enough time reading Darwin, writes Gruber, and you will sense a "modern view of life . . . in which man suffers a double indignity. His future is both determined and unpredictable." In evolution as in a person's life, certainty and chance are intimate companions traveling arm in arm, each momentarily becoming the cause of what follows, emerging with the flap of a butterfly's wings, a mutation, or a tipping point. Darwin mumbled this heresy in a notebook: "I verily believe free will & chance are synonymous. Shake ten thousand grains of sand together one will be uppermost:—so in thoughts one will rise according to law." This figures free will, the keystone of humanity's pride, as its nemesis—natural chance. Darwin was clearly ruthless about traditional distinctions, for all his sweet temper. He played with a homemade deck of cards. Biophilia wrote the rules of his game.

Some contradictions are more complex than others: when two putative opposites, such as chance and certainty, are no longer opposite but one, that's a Möbius strip, the figure-eight strip of twisted paper that appears to have two sides but has only one. The Möbius appears often in Darwin's thought, but it has a precedent in physics. The most famous image of a Möbius strip was created by the graphic artist M. C. Escher, who portrayed ants parading along an infinite loop. You make a Möbius strip by gently twisting one end of a long strip of paper 180 degrees, then taping the two ends together. You may then draw a line on both erstwhile sides without lifting the pencil. If the strip were still a bona-fide two-sided piece of paper, you would have to pick up the pencil to draw a line on both sides. So this quirk of the universe has not one side or two opposite sides but one side, two sides, and two not-opposite not-sides. The Möbius strip is one of those paradoxes of the physical world that inform our formation. Mathematicians will tell you with a straight face that a Möbius strip exists in "one and a half dimensions," which only serves to legitimate bewilderment at this classic topological figure.

The Möbius is a model for Darwin's thought. Gruber described a moment in Darwin's scientific-creative process: "Are we saying that restructuring the argument depends on seeing the principle of natural selection, while at the same time seeing the principle depends on restructuring the argument? Yes. This is a circular argument, or better, a helical process." His process seems to be more of a Möbius, where concept makes sight and sight makes concept recursively.

The Möbius demolishes binaries. When Darwin famously grumbled to himself, "Having proved men's and brutes' bodies on one type: almost superfluous to consider minds," he clearly considered the mind a mere piece of the body, not something on the other side of a binary see-saw. All throughout his evolutionary theory, body and mind are one. Today, caveats about bifurcating the physical and the mental, the body and mind, are issued by other voices. "For those committed to a Cartesian world view, one could think of the eye as a tube that traverses metaphysical realms, one end of which obtrudes onto the physical realm, the other into the mental," wrote the evolutionary psychologists John Tooby and Leda Cosmides. But, they continue, "There is no Cartesian tube; both ends of the visual system are physical and both are mental."

But as Darwin struggled with these complexities and theories, under it all was his confidence in that "one stock," his feeling for all creatures large and small, his "organic beings," "living creatures," and "humble organisms." He took profound pleasure in the ant, the orchid, and Nature's oddest "life forms." All were equally alive, and he loved all equally. He gazed, like a lover, into the heart and soul of nature, into its weirdest contradictions. This is biophilia. Just as the fish are the last to discover the sea, Darwin's biophilia is the sea his thought swam in but did not see.

"THE VERY SUMMIT OF THE ORGANIC SCALE"

Of course the most controversial duality that Darwin demolished but then transformed into continuity was "human" "versus" "animal." This shift was an evolutionary Möbius as well as a cultural rollercoaster. But he asserted the continuity early and often. In one of his first scientific notebooks, he confided: "It is absurd to talk of one animal being higher than another.—we consider those where the intellectual faculties most developed as highest. A bee doubtless would when the instincts were." Scribbled like a true biophile. The lower-higher juxtaposition trivialized the simple creature. As his ideas progressed, he changed his workhorse phrase *organic being* to the locution *organized being*, a promotion of some kind. Darwin scholars have variously wagged their fingers at or lauded him for his suspicions that plants have brains and his awe at the smarts of an earthworm. His prose, wrote the English professor Stanley Hyman, can sometimes resemble "a mad scientist in Hawthorne or Poe." Many clucked at his anthropomorphism—a trait forbidden to the scientist but reflexive empathy for a biophile.

The putative schism between human and animal and their ranking as high and low respectively haunted Darwin as he theorized. A few years after completing *Origin*, he wrote to his friend Sir Joseph Hooker, "with respect to 'highness' and 'lowness' my ideas are only eclectic and not very clear. It appears to me that an unavoidable wish to compare all animals with men . . . as supreme, causes some confusion." Darwin labored to purge his thought of quotidian hierarchy, as when in a book margin he scrawled: "Never use the words 'higher' and 'lower.'" He did, off and on, use these two words. But he declared to Hooker that, regarding the word "higher," "I intend carefully to avoid this expression for I do not think that anyone has a definite idea what is meant by higher."

One sign of Darwin's stance was how often he damned *Homo sapiens* with faint praise, as in *The Descent of Man*. He equivocated that "Man . . . is the most dominant animal that has ever appeared on the earth" and that our species has "risen . . . to the very summit of the organic scale." Sometimes he did not pull his punches: "Man [has] a pedigree of prodigious length, but not, it may be said, of noble quality." And he restates in an admonishment to taxonomists that "man has no just right to form a separate Order for his own reception."

There was more than sheer gall in Darwin's avowing continuities between animals and *Homo sapiens* and in his classifying humanity as just one more animal. If "Man," his mind, and his "highest faculties" were merely extensions of animal faculties, that knocked humanity off its pedestal as the image of a divine creator. Darwin addressed this point in *Descent*, blunt as blunt could be about the continuity between amoebas and apes and us, when he ended this lurid scientific tale with a recap of the bad news: "Man, with all his noble qualities, with sympathy which feels for the most debased, with benevolence which extends not only to other men but to the humblest living creature, with his god-like intellect which has penetrated into the movements and constitution of the solar system—with all these exalted powers—Man still bears in his bodily frame the indelible stamp of his lowly origin." The naked truth about the naked ape, and he managed to use the word "god," covertly bringing in the crux of the matter. But in the final sentence of this book, Darwin invokes the human as having "benevolence which extends . . . to the humblest living creature." That is as good a declaration of biophilia as any.

WHOM TO TRUST?

Biophilia resembles religious worship, according to a dusty scholar quoted by Williams James in *The Varieties of Religious Experience*: "Not God, but life, more life . . . is . . . the end of religion. The love of life, at any and every level of development, is the religious impulse," asserted James H. Leuba. Darwin might agree. Darwin's staunch ally, the prominent scientist Thomas Henry Huxley, conflated not God and the love of life but religion and evolution. Huxley heralded the "Hegira of Science" exiting "from the idolatries of special creation" and heading toward "the purer faith of Evolution." Peirce idolized evolution. Darwin did not worship evolution, but he glorified its "grandeur." He revered the pulse of life "at any and every level of development." He had no objections to reverence per se. He said in *Descent* that the human "feeling of religious devotion" descends from "the deep love of a dog for his master," making veneration most natural. And though Darwin was once "tempted to believe phrenologists," if the subject at issue was devotion, he would have taken the word of a dog over that of a phrenologist any day.

Chapter Two

IT'S DOGGED AS DOES IT

I started early, took my dog—

—Emily Dickinson

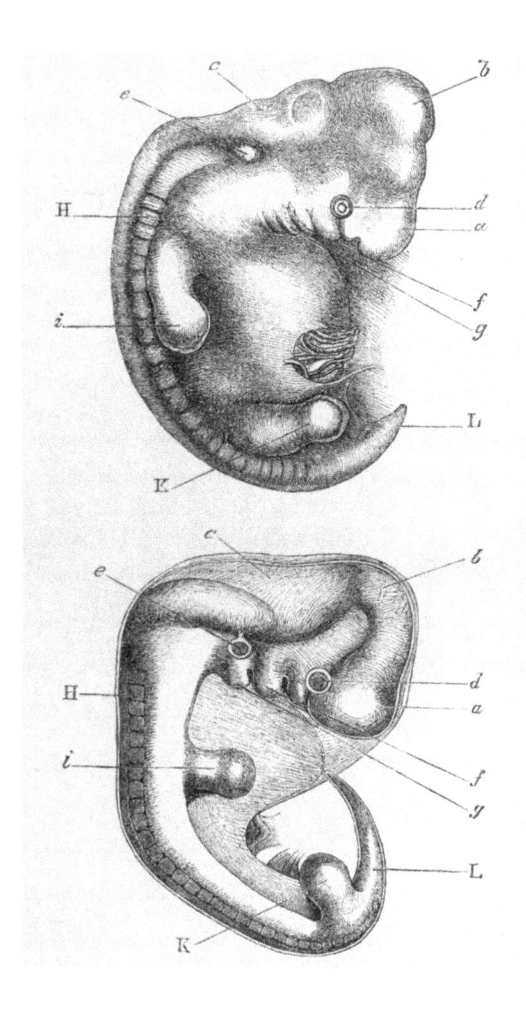
c
b
e
H
d
a
f
i
g
L
K
c
b
e
H
d
a
i
f
g
L
K

FIGURE 2.1 A tale of two embryos.

This comparison of a human (top) and dog in utero is the first illustration Darwin offered in *The Descent of Man*, declaring that the human "embryo itself at a very early period can hardly be distinguished from that of other members of the vertebrate kingdom." As if one small diagram could prove the truth of his evolutionary theories for once and for all, he went on: "It would be superfluous on my part to give a number of borrowed details," though he had many to hand. On matters ranging from the dog's physical structure to its moral backbone and love of play, Darwin had recorded thousands of facts and observations. In his books, he continually resorted to the canine to show similarities and continuities across species, evidence of evolution. He wrote anthropomorphically, almost unscientifically, about dogs time and time again, and the most dog-besotted anecdotal evidence was telling to him. Darwin was, in fact, a sentimental sap about dogs. His life was populated by dozens—during his childhood and college years, while circling the world, and in the home he eventually built with his wife. The dog stories in Darwin's life sparkle with the usual joys, shimmer with the usual tears, and light up the ordinary depths of canine companionship. But all glow with the radiant passion of the man who saw "one living spirit" in the myriad forms of organic beings.

Source: Reproduced with permission from John van Wyhe, ed., *The Complete Work of Charles Darwin Online*, 2002–, http://darwin-online.org.uk/.

"POLLY DIED," Emma recorded in her diary on Thursday, April 20, 1882. That was the day after Darwin himself died. Polly was a terrier mix who first belonged to their daughter, Henrietta. When Henrietta married, she entrusted Polly to her father, as others had done with their superseded dogs. The grand man of science doted on Polly, according to Francis's memoir, and taught her parlor tricks. One was: Darwin would place a "biscuit" on her nose, and she would await his cue for her to toss and gobble it. Huxley nicknamed Polly "the Ur-hund" to razz his reclusive friend about his pet. Emma wrote to her daughter that Polly became "perfectly devoted" to Darwin after a litter was taken from her; she speculated that Polly had adopted him as a substitute puppy. Polly licked his hands with an "insatiable passion," he wrote in *The Expression of the Emotions in Man and Animals*. She followed him underfoot, as certain as a shadow in a desert. Days she spent in a pillowed basket on the hearth in his study. Nights Emma had to "drag her away" from his side, she told Henrietta.

The poet Rainer Maria Rilke chronicled his "glorious" feeling when sustaining eye contact with a dog. He invented a verb for it, translated as "to insee . . . to let oneself precisely into the dog's very center, the point from which it becomes a dog, the place where God, as it were, would have sat down for a moment when the dog was finished." Rilke's communion must have been familiar to Darwin. What is certain is that when Darwin suffered the heart attack that proved fatal, Polly "became very ill with a swelling in her throat . . . creeping away several times as if to die," Francis scribbled. The canine urge to suttee is the stuff of legend, of course. Non-dog-people will see as the cold cause of Polly's death a happenstance illness or Francis's consequent act of euthanasia. But "the word 'cause' is an altar to an unknown god; an empty pedestal," William James told us, and dog people will say Polly died of heartbreak.

Even as Darwin pondered the blush of the octopus, the feather of the peacock, and the sex life of the barnacle, he was fascinated by the yawn of the dog. His mind touched it like a talisman. "Seeing a dog & horse & man yawn, makes me feel how much all animals are built on one structure," he confided in a notebook. The yawn is ubiquitous and quirky, ridiculous and mechanical, disgusting and luxurious, unpredictable and uninterruptable, its purpose a mystery even today. He saw the yawn as a synecdoche of the "one structure" just six short sentences after shoe-pounding in his notebook: "Origin of man now proved. He who understands baboon would do more toward metaphysics than Locke." The ape may be the first animal we think of when we think of Darwin, but the dog was ever a dousing rod that led him to home truths and to see *Homo sapiens* as just one more organic being formed from the "one structure." The dog was ever his first example of love, joy, reasoning, and virtue. Dogs are a veritable apogee of biophilia. And when Darwin studies dogs, passion and perception flow from his biophilia.

PRANKS AND HUNTING

The child Charles was the father of the observant experimentalist. And dogs, as he commented decades later in *The Descent of Man*, offered "the best opportunity for observation." Francis recalled one of his father's boyhood stories, about playing hide-and-seek with a hunting dog. Darwin would ask a confederate to cloister the dog; then he would run around a huge garden, laying a track, and then climb a tree. The dog would be released into the garden, giving Charles "the fun of watching it" search out his scent, sniff its way to his tree, and discover him perching, providing clear evidence of its tracking skills. As a child Charles also understood the heart of dogs. He was "an adept in robbing their love from their masters," he boasted in his otherwise modest autobiography. Growing up, he repeatedly stole the affections of his sister's dogs—family lore that no doubt eased Henrietta's decision to leave Polly in his care. In college, he reprised this stunt: living in the same residence hall as his cousin, he beguiled his cousin's dog to sneak out of its master's rooms at night to sleep under Charles's blankets, at his feet.

This deep connection with nonhuman animals is fundamental in biophilia. It is a quality that runs in families. Charles's brother, Erasmus Darwin, who socialized with glitterati such as the historian Thomas Carlyle, was admired by Carlyle's wife for his way with dogs. A Darwin cousin, Francis Sacheverel Darwin, wrote a book called *Gamekeeper' s Manual* that was praised by Charles's son for its "keen observation of the habits of various animals." But Charles could win the loyalty of even a misanthropic dog.

Darwin's father and paternal grandfather were quite clearly endowed with biophilia. Both excelled in diverse professions, indicating a deep engagement with life, an overarching acuity. Darwin's grandfather Erasmus practiced medicine with much success, advanced experimental science, and had a fact-inspired imagination. In an 1883 eulogy on Darwin delivered to the Leeds Philosophical and Literary Society, one Louis Compton Miall described Erasmus as "a mechanician, a naturalist, and a poet . . . famed for his inventions: all intended to abridge labour or serve mankind. . . . He contrived a horizontal windmill . . . ; a knitting loom; a weighing machine; a flying bird; a canal lock; a rotatory pump; wheels with elastic spokes. . . . He had a speaking-tube put up in his house, to convey messages to the kitchen." Erasmus published "The Loves of the Plants," a biophile's treatise simultaneously technical and raunchy, and his domestic life was a scandal. In his love of life, he walked on the wild side. Erasmus shared visions with the "lunarticks" of the "Lunar Circle"—a group of inventors, capitalists, aristocrats, and academics, all fascinated by natural philosophy, who first held meetings only when the moon was full, as traveling was easier with light. The group hosted the likes of Benjamin Franklin as speakers. Today, the Lunar Circle is the stuff of legend for scientists and techie entrepreneurs. Erasmus was brimming with biophilia.

Darwin's father, Robert, was a gentleman farmer, an astute financial investor, and a reputed country doctor. Darwin attributed to his father an "intuitive perception of character" and a "most remarkable power . . . of reading the characters, and even the thoughts of those whom he saw even for a short time. We had many instances of this power, some of which seemed almost supernatural." Robert's nose for signs of hidden truth in patients was a diagnostic aid one cannot patent. Darwin turned this same insight on all living organisms. The Nobel laureate and biologist Albert Szent-Györgyi quipped, "Discovery consists of seeing what everybody has seen and thinking what nobody has thought." Which only raises the question of why just that one person thinks what nobody else has thought. Perhaps in Darwin's case, the answer is Darwin's natural-born but singular case of biophilia.

It is a truism today that even the scientist's perceptions are, willy-nilly, filtered by ideas. For example, the Darwin scholar Dos Ospovat has written, "Even when Darwin was in his closest contact with nature, as on the voyage and during his research on barnacles, his interaction with nature was mediated by assumptions and ways of perceiving nature that he derived from other naturalists . . . and from the culture in which he was educated and carried out his work." Yes and no. Because notwithstanding his intimacy with barnacles, Darwin's closest contact with nature was when he was a child, when theory had no toehold in his mind, when logic would not filter perception, when a puppy love for life begins. When he was a child, no a priori "assumptions" nor well-thought-out philosophies "mediated" his interaction with nature. Nor can culture determine a child's love for the nonhuman organic being: within any one culture, attitudes toward, say, dogs vary so widely, a child might choose which to "derive"—a drive to dominate? a fear of the other? the bond of biophilia?

Darwin's communion with dogs deepened in his teens. He learned to hunt and to deploy dogs in time-honored ways. The hunting season of the English leisure class was the glory of many years for him, until his father, annoyed with his lackluster progress in college, accused him: "You care for nothing but shooting, and dogs, & rat-catching, & you will be a disgrace to yourself & all your family." A doozie in the annals of mistaken parental prognostications. But whether or not we agree that Charles dragged the Darwin name into the mud, Robert got it right about his son's love of dogs.

Those testosterone days weighed on Darwin's conscience, and he mused about them in the published version of the autobiography. "I think I must have been ashamed of my zeal for I tried to persuade myself that shooting was almost an intellectual enjoyment; it required so much skill to judge where to find most game and hunt the dogs well." He soon ceased senseless killing for sport, and in later decades, he felt guilt at inflicting suffering. But however benighted the pursuit, hunting did give him a wealth of knowledge. It expanded his "affinity for life," in the phrase of Wilson.

And however attenuated the English hunt was qua hunt, Darwin acquired enough outdoorsmanship to hold his own among men for whom hunting skills were tantamount to surviving. When he traveled around the world on the British government's mapping expedition, such men guided him in rough wildernesses. He trekked across South America, for example, escorted by Indigenous guides; they threaded their way through forests, galloped across plains, and slogged up and down mountains and ravines, always with dogs. Then and throughout the five-year voyage, Darwin contemplated the presence, absence, and character of dogs.

Darwin's biophile connection to animals aided his bond with his escorts. In his bestseller, *Journal of Researches Into the Natural History and Geology of the Various Countries Visited During the Voyage Round the World of the H.M.S.* Beagle (Darwin did not dub the ship but delighted in joking about its namesake), Darwin described the gaucho's lasso and its cousin, the bolas, a thong weighted with balls at its ends. Thrown in a whirl at prey, the bolas twines around its target in a stranglehold. He detailed his buffoonery in learning to wield this tool:

> The main difficulty . . . is to ride so well, as to be able at full speed, and while suddenly turning about, to whirl them so steadily around the head, as to take aim: on foot any person would soon learn the art. One day, as I was amusing myself by galloping and whirling the balls round my head, by accident the free one struck a bush; and its revolving motion being thus destroyed, it immediately fell to the ground, and like magic caught one hind leg of my horse; the other ball was then jerked out of my hand, and the horse fairly secured. Luckily he was an old practised animal, and knew what it meant; otherwise he would probably have kicked till he had thrown himself down. The Gauchoes roared with laughter; they cried they had seen every sort of animal caught, but had never before seen a man caught by himself.

Darwin risked and reaped the ridicule of his employees. He was always easy-going, never self-important, and likely effaced his position as employer. More to the point, I think, he felt sure of their respect. As they journeyed, his guides must have seen his savvy in the primal arts: his feeling for horses, command of dogs, sharp marksmanship, and willingness to dine on the day's takings—be it rhea (called an ostrich in Darwin's day) eggs or a puma. He embraced the "adventure of living," as in Fromm's description of biophilia. These traits must have counterbalanced the klutziness of the rube, his malodorous colonial origins, and the congeries of outlandish tools and scientific protocols he used to collect his thousands of specimens of flora and fauna, fossil and rock. As the obscure editor Frans Verdoorn noted, Darwin's *Journal of Researches* "gives a revelation of a vigorous and robust individual which goes far toward explaining his later achievement." There is irony in Darwin's being belittled by his father as a dilettante in his own culture, when he was practicing arts essential in another—and essential also to his work as a naturalist. That is three cultures clashing at one nexus. The anecdote also highlights Darwin's love of the new and his high spirits, which is to say, his biophilia at full tilt.

THE LION IN HIS DEN

Darwin's return to England was followed by a few dogless years. This lack may explain why, as he began to jot down the pros and cons of marrying Emma, among the pros was a grumbled "better than a dog anyhow." Decades later, he mentioned in a book that among the Turama Indians, men paid the same price for brides and "good" dogs. But his evaluative exercise ended by proving marriage logical, so much so that he scrawled QED below the dueling columns. Emma's reasons for accepting his proposal included that he was "humane to animals." Emma herself was an "adept" at reading dogs, to borrow Darwin's word. Emma's simpatico biophilia boded well for their union (see chapter 7). Over the course of forty-three years of marriage, the Darwins had at least twenty-one dogs in their home.

As Charles and Emma domesticated their new home, Down House, he stashed thoughts in notebooks, like a puppy burying treasures in the woods. He foraged among ideas. He drew from tracts on philosophy and husbandry and from letters from his friends. In all, he eventually recorded and pondered thousands of facts, anecdotes, and musings on the dog's breeding, conscience, habits, instincts, imitation, logic, magnanimity, modesty, mourning behavior, pride, selection by breeders, training, trickery, value in human survival, and more. He did not neglect to mention the dreaming of dogs, which showed "similarity of mind." When he was writing *The Variation of Animals and Plants Under Domestication*, which contained his data for *On the Origin of Species*, he complained in a letter: "You would be surprised how long it took me to pick out what seemed useful about dogs out of multitudes of details."

Darwin saw many likenesses between *Homo sapiens* and best friend. Former houndsman and incorrigible theorist that he was, Darwin tracked commonalities in the wilds of his biophile mind, cozy in his study, with the company of one dog or another. Beyond the yawn, other resemblances Darwin noted were pleasure taken in a good tickle and the likeness of a dog's snarl and a man's "lip . . . stiffening over his canine teeth." Laughing, he declared, is "modified barking." Other similarities: both "speak in their sleep" and bite during sex. Common emotions were common to human and dog, such as compassion, courage, homesickness, and triumph. Canine parental and sexual love seemed hominid as he wrote about it, and vice versa.

Darwin mined his files as he assembled his later theories. The notebook comment on dreaming, for example, he plumped up decades later in *The Descent of Man*: "As dogs, cats, horses, and probably all the higher animals, even birds . . . have vivid dreams, and this is shewn by their movements and voice, we must admit that they possess some power of imagination." Note his rhetorical "we must admit," addressing his readers' potential reluctance while emphasizing the need to see facts (see chapter 3). Between the lines, Darwin acknowledged skepticism, as if he knew some readers might find implausible the idea of a dog having an imagination, as if he knew some humans see other animals as alien, as if he knew that not everyone has the same perception of oneness, the same understanding informed by biophilia.

For Darwin, language is a significant commonality shared by dogs and humanity. Language was a provocative subject: That language is exclusive to humans was a sacred cow in arguments saying we are in god's image and thus entirely distinct from animals. When introducing language as a topic in *Descent*, Darwin recognizes this point of view, noting that language is "one of the chief distinctions between man and the lower animals" in the thought of his day. But thirty years before, in a private notebook, he had cautioned himself not to "overrate the distinction." He accorded linguistic prowess to animals because he considered language broadly, as specific sounds that communicate specific meaning. By the time he sat down to write *Descent*, he had found an archbishop—science makes strange bedfellows too—to speak for his view: "But man, as a highly competent judge, Archbishop Whatley, remarks, 'is not the only animal that can make use of language to express what is passing in his mind, and understand, more or less, what is expressed by another.'"

Darwin pointed out that the canine ability to communicate increased as dogs wormed their way into the role of best friend. We might "suspect," he wrote, that the dog's various barks result from their "having long lived in strict association with so loquacious an animal as man." He hawked the dogs' diverse vocalizations as a "new art" and hailed it a "remarkable fact that the dog, since being domesticated, has learnt to bark in at least four or five distinct tones." The "remarkable fact" is proof for him that other species learn and deploy language. "That which distinguishes man from the lower animals is not the understanding of articulate sounds, for, as everyone knows, dogs understand many words and sentences." "Everyone knows?" Everyone does not include the many people terrified of dogs. This is Darwin's biophilia, along with his dog-centric mind, controlling his pen.

In his notebooks, he observed that dogs understand human facial expressions. Dog owners know well the ability of dogs to read minds and know mood. According to Darwin, in fact, this kind of nonverbal communication is more significant than language. In *Descent* he writes, "Man . . . uses, in common with the lower animals, inarticulate cries to express his meaning. . . . This especially holds good with the more simple and vivid feelings. . . . Our cries of pain, fear, surprise, anger . . . the murmur of a mother to her beloved child, are more expressive than any words." So much for the hegemony of the articulate. Even the sophisticated syntax of human language, he asserted in

Descent, reflects only an evolutionary change in degree, not a change in kind. One way or another, in Darwin's hands, language goes to the dogs.

Darwin also saw the origins of religion in the behavior of the dog. In *Descent*, he pointed out that "primitive" religions—what today we might call Indigenous or pagan—are often animistic, in accord with a "tendency in savages to imagine that natural objects and agencies are animated by spiritual or living essences." He speculated that the dog imagines likewise, as is "perhaps illustrated by . . . my dog, a full-grown and very sensible animal . . . [who] was lying on the lawn during a hot and still day; but at a little distance a slight breeze occasionally moved an open parasol. . . . Every time that the parasol slightly moved, the dog growled fiercely and barked. He must, I think, have reasoned to himself in a rapid and unconscious manner, that movement without any apparent cause indicated the presence of some strange living agent, and no stranger had a right to be on his territory."

As an aside, note how he presented the dog's reasoning and consciousness. In detailing his observations, he appoints himself interpreter. Our usually logical thinker makes not a few anthropomorphic assumptions here as he voices the dog's logical chain. While Darwin has often been chastised for his anthropomorphism, today, in fact, his approach is in an analytic discipline scientists call "anecdotal cognitivism." Or his anthropomorphism might be understood as a result of the oneness of biophilia.

He proceeded to compare the "feeling of religious devotion" in "civilised man" to a feeling in dogs, opining that "we see some distant approach to this state of mind, in the deep love of a dog for his master, associated with compete submission, some fear, and perhaps some other feelings." He then handed the punch line to another authority, thereby offloading the heresy: "Professor Braubach goes so far as to maintain that a dog looks on his master as a god." That may or may not be true, but for Darwin, the dog was certainly a good soul. And even if Judeo-Christian *Homo sapiens* believe that their God handed down the Golden Rule, "Do unto others as you would have them do unto you," Darwin knew that dictum to be older than culture; he scribbled in an early notebook, "moral sense may arise from strong instinctive sexual, parental, & social instinct, giving rise [to] 'do unto others as yourself.'" In his view, *Homo sapiens* arrived at the Golden Rule because our soul was formed not in God's image but the dog's. Kindness is innate in the biophilia of dog and human.

DOGGED DOG, DOGGED MAN

As Darwin churned out theory after theory, study after study, he became a stay-at-home country squire, a venerable bump on a log. He became ever more an invalid over the years but took walks almost daily to think and exercise, usually with a dog. For years, this was the surly Bob, intransigent to all other humans but fanatical about jaunts with Darwin. As Darwin aged, the sprite Polly monitored him on his walks, sometimes enduring a pelting rain to stay at his side. Amid Darwin's voluminous correspondence with eminent scientists the world over, he found time to reply to the local vicar, who shared an adorable bad-dog story: "Thanks for the very curious story about the dog & mutton chops. They are wonderful animals, & deserve to be loved with all one's heart, even when they do steal mutton chops."

Petty thieves notwithstanding, Darwin imitated the dog in his research. Sheer "doggedness," according to his son, spurred those decades of experimentation in his laboratories. He was assiduous, working with scrupulous records, nimble fingers, skinflint efficiencies, and an unending eagerness for the morning's news from the bench. The engine for his productivity was, Francis wrote, his "power of sticking to a subject; he used almost to apologise for his patience, saying that he could not bear to be beaten, as if this were rather a sign of weakness on his part. He often quoted the saying, 'It's dogged as does it,' and I think doggedness expresses his frame of mind almost better than perseverance. Perseverance seems hardly to express his almost fierce desire to force the truth to reveal itself." A pointer has a fierce desire to reveal the bird, the herding dog has a fierce desire to chase off the coyotes, and Darwin had biophilia's fierce desire for learning truths and facts.

Polly did right to die in his honor.

Chapter Three

THE GLORIES AND LIMITS OF FACTS

The situation had changed by—well, by whatever there was, by the outbreak of the definite.

—Henry James

FIGURE 3.1 The orchid and the moth.

In one of those true stories that become legendary, Darwin predicted the existence of a moth like this one—sporting a twelve-inch tongue. Darwin based his 1862 prediction on the Madagascan star orchid's twelve-inch nectary—a dangling, hollow tube with nectar pooled at its base. Darwin's sense of coevolution told him that some moth must have evolved a matching proboscis to procure that nectar. "This belief of mine has been ridiculed by some entomologists," Darwin wrote, even as this very same hypothetical moth-orchid duo was considered proof of the existence of God by the Duke of Argyll in his 1867 book *The Reign of Law*. Darwin's colleague and sometime rival Alfred Russel Wallace was persuaded, however, that such a moth existed, and he asked the natural history illustrator Thomas William Wood to prepare this image for his 1867 article "Creation by Law." In a footnote to the article, Wallace exhorted: "Naturalists . . . should search for it with as much confidence as astronomers searched for the planet Neptune,—and they will be equally successful!" They were. In 1903, Karl Jordan and Lord Walter Rothschild discovered a new sphinx moth, with the predicted proboscis, providing yet another proof of Darwin's preternatural capacity to intuit facts. In 2019, a children's book was published on this fabled tale in science.

Source: Reproduced by kind permission of the Syndics of Cambridge University Library. Photograph, *Quarterly Journal of Science*, PR-Q-00340-00001-C-00007-00004-000-p.471.jpg.

SOME FACTS are quotidian, some lurid, some elusive. Drifting through college, Darwin often took a naturalist's walks, collecting as he had since childhood. Such collecting was a fashion among the European men of his day. Darwin collected catholicly: plants, minerals, bones, rocks, crawling critters, the usual and the unusual. His collections while in college earned plaudits from his professors, and he pursued such extra credit opportunities because he was a lousy student. One day, when he was trolling in the local marshes, he "saw two rare beetles and seized one in each hand; then I saw a third and new kind, which I could not bear to lose, so that I popped the one which I held in my right hand into my mouth. Alas it ejected some intensely acrid fluid, which burnt my tongue so that I was forced to spit the beetle out, which was lost." He told this story as "proof of my zeal." He wanted to capture the rare beetle, of course, and to name a new organic being, but the zeal was for new species and new facts. All his life, facts were prey for Darwin. His biophilia needed them, to return to the touchstone of Fromm, "to construct," to fuel and satisfy his "wondering," and to fulfill his desire "to see something new." And then too, the discoverer dwells in what Fromm called "the adventure of living"—and discovery of the elusive is most certainly an adventure.

New facts are, to borrow from Henry James, "the outbreak of the definite." Intrinsic to biophilia's love of life is its seeking the outbreak of the definite, a drive for truth, with truth in this case the facts that comprise life. The love of fact is an aspect of the love of nature—"attracted by the process of life and growth," in Fromm's words, along with the attraction for the new and the true about how life lifes. Discovering facts is, in one way, the ne plus ultra of a scientist's biophilia.

In his passion for knowledge, Darwin chased facts as intrepidly as if they were rare and possibly noxious beetles. He ran experiments and made observations at home for decades, in his greenhouses, observing whatever was to hand, from earthworms to his infant children. Always hungry for more information, he was a fastidious and inventive experimentalist, a diligent morphologist, and an astute taxonomist. His love of facts was such that he was able to discern that "fact" is a genus with many species. He distinguished among humdrum facts, hard-won facts, false facts, missing facts, counterintuitive facts, unproven facts, unknown facts, and unknowable facts. His biophilia gave him a feeling for factuality.

FUN WITH FACTS

One of Darwin's favorite tools as a writer was the colorful fact. In *The Descent of Man and Selection in Relation to Sex*, he is generous with these. He parades a motley crew of drunken monkeys; males who breastfeed, including the occasional *Homo sapiens*; vengeful animals; and people who wiggle their ears. He introduces cultures that honor "lascivious" women. He presents laboratory work comparing far-flung specimens of lice. Why so lurid? To make it easy for scientists and society to squelch his book? Many of his peers, and another generation or two of scientists, viewed Darwin as dotty when he wrote *The Descent of Man*. Nonetheless, as the unofficial sequel to *On the Origin of*

Species, the book was an instant bestseller. He remarked, "I suppose abuse is as good as praise for selling a book." The purple details signal that *Descent* is an exposé on the scandalous origin of *Homo sapiens*; they show a Darwin using facts for shock value, telegraphing to his readers, "Heads up!"

Darwin could not refrain from flashing weird facts. Here he is rhapsodizing about a larval barnacle, "with six pairs of beautifully constructed natatory [for swimming] legs, a pair of magnificent compound eyes, and extremely complex antennae." Francis reminisced, "We used to laugh at him for this sentence, which we compared to an advertisement." Darwin had, Francis shrugged, a "tendency to give himself up to the enthusiastic turn of his thought, without fear of being ludicrous." His thoughts continually turning, of course, to the miraculous facts of nature. Baby barnacles, orchids, eyes, yawns, carnivorous plants. Every datum he examined was one of a myriad of "splendid decorations," "marvellous structures," "ingenious contrivances," and "extreme beauty." He was prone to superlatives.

In *Descent*, he revealed, making no bones about it, that we are born of an arboreal quadruped who sported a tail, a pelt, long teeth, pointy ears, and prehensile feet. Each factoid added insult to injury. He also offered his guess that a tiny hermaphroditic sea creature was the ancestor of the ancestral quadruped. His bold assertion has of course since been proven; when Darwin speculated, he was usually elucidating a fact of the as-yet-unproven variety, a specialty of his. He was compelled, of course, to broadcast the news about the ancestral aquatic hermaphrodite, and this was likewise bold, given that eighteenth-century Western botanists refused to accept that hermaphroditic plants were normal.

His son Leonard wrote an anecdote about Darwin's capacity for wonder at the wonders of life: "In company with Dr. Ogle, a keen student of evolution, he [Darwin] was wandering about the garden when he paused to pick some flower, and then said that it was staggering to have to believe that the beautiful adaptation which it showed was the result of natural selection. To this Dr. Ogle quietly replied, 'My dear sir, allow me to advise you to read a book called *The Origin of Species*.'" Darwin's awe at nature was integral to his modus operandi, to his charm, and to a biophilia that reveled in life's facts, the odder the better.

FACTS SERVED WITH SUGAR

With the empathy of the oneness of biophilia, Darwin was able to imagine the needs of his readers. *On the Origin of Species* was a bestseller—as was his travelogue on his trip around the world—and it remains in print today in innumerable editions. Why his version of evolution and not others? Because he was right, of course, but also because he was a good writer. Along with grabbing his readers' attention with sensational details, he crafted prose about his facts that appealed to the imagination. Many scholars have pointed to Darwin's lucid, concise, and at times charming style of writing, which certainly eased the entrance, if not the embrace, of evolutionary facts into the public mind. As noted earlier, he often began his explanations by modeling his own ignorance. This tactic is of a piece with his modesty, of course. But it is also pragmatic: When he tells of his own intellectual enlightenment, he invites his readers to join him on similar hegiras of the imagination. He assuages his readers' nervousness about new information. Then, detail by slow detail, idea by stepwise idea, he launches the reader into the process of discovery and understanding. He connected readers to his ideas by, ironically, reminding them that he himself had once lacked facts and understanding. He strengthened his case by admitting confusion, a ploy most would not try.

Another tool Darwin used to forward facts is to model his readers' skepticism: "It was a long time before I saw how." Francis commented, "His courteous and conciliatory tone toward his reader is remarkable. . . . The reader is never scorned for any amount of doubt which he may be imagined to feel, and his scepticism is treated with patient respect. A sceptical reader, or perhaps even an unreasonable reader, seems to have been generally present to his thoughts." For example, early in *Origin*, Darwin described the domestic animals that breeders had shaped for various traits, using the term "artificial selection." He confessed to his own astonishment at the diversity achieved by breeders: "I have discussed the probable origin of domestic pigeons at some length; because when I first kept pigeons . . . I felt fully as much difficulty in believing that . . . they had all proceeded from a common parent, as any naturalist could." Acknowledging his skepticism is tantamount to promising his sympathy. And indeed, Darwin explicitly hoped his facts would be helpful: "This accompanying diagram will aid us in understanding this rather perplexing subject."

Early in *Descent*, he invited inquiring minds to join him in a thought experiment: "He who wishes to decide whether man is the modified descendent of some pre-existing form, would probably first enquire . . ." He complimented his readers' open minds.

He used empathy as a rhetorical trope. He gave a spoonful of sugar to help his readers swallow complicated facts. He positioned himself as there to help, the whole rough, long, conceptual, philosophical, spiritual, contradictory, biological way. How can we not appreciate that he offers to "aid us in understanding"? We cannot help but return his empathy by entrusting him with our ignorance, this man who is possessed by a question, has struggled for and with an answer, and attends to our confusions.

Darwin even understood that not everyone is as enamored as he is of the infinity of facts, just as he always understood opposite viewpoints. He told readers of *Origin*, "To treat the subject properly, a long catalogue of dry facts ought to be given; but these I shall reserve for a future work," and indeed that material filled another large volume, *The Variation of Animals and Plants Under Domestication*. On one occasion, Darwin invited his readers to skip most of his book: early on in *The Power of Movement in Plants* he wrote, "No one who is not investigating the present subject need read all the details, which, however, we thought it advisable to give. To save the reader trouble, the conclusions . . . have been printed in large type. . . . He may if he thinks fit, read the last chapter first, as it includes a summary of the whole volume; and he will then see what points interest him, and on which he requires full evidence." Darwin arranged a choice, a leisurely and lush stroll for factophiles or just the truth in a nutshell for the efficient. He expressed a oneness with all his readers.

MYSTERIES

The flip side of a biophile's joy in facts is an obsession with mysteries—which are, quite simply, missing facts, hidden truths to be discovered and proved. A biophile such as Darwin is ever driven to achieve the outbreak of the definite. Not everyone is brave enough to chase the unknown. Darwin's particular factophilia enabled him to use the known to build his theories, the unknowns. He opened *Descent* by telling readers that the book contained no new facts. This move was not scientific modesty but the honest truth; the book was all about theory. The modern-day evolutionary theorist John Tyler Bonner likewise prefaced his classic book on the evolution of culture in animals by noting that there was nothing new in his book; he had just put together the facts of others in a new way. Both theorists then proceeded in pure speculation, grinding new theoretical lenses.

But Darwin knew that even with his own carefully wrought lenses, he could not see everything. In his theories he left niches for the unexplained, placeholders for mysteries. When confronted with the lack of a reason or knowledge, he did not get upset, jump to conclusions, or resort to preexisting certainties. And lack of a complete explanation did not invalidate a theory for Darwin. Quite the contrary, as he sought theories based on facts he trusted, he held in mind a question that served as a North Star for his intellectual odyssey: What don't we know?

When he noticed an unanswered question, he would confront it. "In what manner the mental powers were first developed in the lowest organisms, is as hopeless an enquiry as how life first originated. These are problems for the distant future, if they are ever to be solved by man." Ignorance is inevitable. His books offer caveats like: "our knowledge, imperfect though it may be." As noted already, he wrote of "Nature . . . We are too blind to understand her meaning." If Darwin lacked an answer, that did not mean he was wrong; it meant only that he lacked an answer. He had faith that some day a fact would nestle into the space he had allocated to the mystery. So when he lacked a fact, when he "failed to obtain clear evidence," he did not let that lack stop him. He might trust his logic or his intuitive sense of the biological truth and then head on to expound a theory eventually confirmed.

One fact without explanation that once obsessed him concerned dogs. He described in *Descent* the behavior of mushing dogs who "diverged and separated when they came to thin ice, so that their weight might be more evenly distributed. This was often the first warning and notice which the travelers received that the ice was becoming thin and dangerous." Science still does not know how the dogs know, what questions mushing dogs ask of the ice. Do their paws register a certain vibration, do their ears hear a different sound in their footfalls, do they scent fish, do they see a distinctive hue under the surface? Darwin asked himself, is their response instinct or intelligence? Then he shrugged: "Questions of this kind are most difficult to answer." But he knew that dogs perceive what we cannot, and they know what we cannot. He accepted our relative ignorance, ever humble, ever ready to acknowledge mystery, and, of course, ever ready to admire the dog.

Mystery justifies Darwin's lifelong epistemological stance: If we understand that there is something we do not know—how the dogs sense thin ice—then we must wonder what else we do not see and what else we do not know. How many other organic beings see and know still other facts that the human sensory equipage is deaf, dumb, and blind to? The modern ethologist Temple Grandin points out that "the reason we've managed to live with animals all these years without noticing many of their special talents is simple: we can't see them." This syncs with Darwin's reflexive acceptance of mysteries. In the same vein, he asked himself, also famously, what a bee might think of *Homo sapiens*' self-nomination as a superior species.

Like the renowned twentieth-century physicist Werner Heisenberg decades later, Darwin knew that our technē for apprehending real-world phenomena may be distortive and insufficient. Heisenberg warned us that in the realm of particles, the act of measuring may alter the physical behavior of what is putatively being measured. He toppled measurement as a metric of truth. He demolished certainty. Darwin was intimate with uncertainty. He prefigured Heisenberg when he cautioned us that it is "always advisable to perceive clearly our own ignorance."

Darwin knew very well that what we call facts at any given moment are not necessarily proof of anything and are no guarantee against ignorance. Facts could be misleading, incomplete, ambiguous, elusive, misinterpreted by ignorance or prejudice, or "discovered" with faulty tools. Facts are never the whole story. He was cognizant of knowledge as a patchwork of assumption and verification, theory and observation. The modern evolutionary biologist George Williams wrote, "One of the strengths of scientific inquiry is that it can progress with any mixture of empiricism, intuition, and formal theory." Darwin exactly.

He had blind faith in the logic of life, even when eye and technē could not see it. He tendered the "supposition that the geological record is far more imperfect than most geologists believe." Or he might conclude sadly, "We see still more plainly the obscurity of our subject," a comment that also reveals his sense of paradox. Science must march forward as best it can even without a phalanx of soldiering facts, is his message. Perhaps this attitude explains why Darwin was nonplussed by a large hole in evolutionary theory, the infamous "missing link" between apes and *Homo sapiens.* The nineteenth-century English poet John Keats described a "man of achievement" as "capable of being in uncertainties, mysteries, doubts, without an irritable reaching after fact and reason." Just so, Darwin was at ease with the unknown, even as he drove to know fact and reason. The unknown was his business, after all. And in his biophilia, he accepted that Nature was always going to be smarter, stronger, and swifter than any old naturalist, himself included.

One mystery Darwin did solve could have been dubbed "The Riddle of the Peacock's Tail" by his early fan Sir Arthur Conan Doyle. How the peacock's tail signaled to Darwin the theory of sexual selection is now part of Darwin's legend. It had been a mystery with a powerful pull for him, and one of his most-quoted comments is an exasperated remark in a letter: "The sight of a feather in a peacock's tail, whenever I gaze at it, makes me sick." Why gaze, then? The answer, for Darwin: the biophile's urge to wrestle down mystery. The burning question for Darwin, who had hammered out natural selection as favoring the well adapted, was: Why does the peacock have a sweeping and stunning plumage so cumbersome it hampers his survival when hunted? The extravagant plumage is a poor adaptation if the goal is survival of the fittest. Moreover, the peacock's female conspecific, the peahen, sports only dull and safe stubs; why do the two sexes have two different features—one ostensibly maladaptive—if they face the same predators?

His answer (eventually proven true) was that evolution gave females an instinctive preference for the most beautiful birds, and such beauty, as we know now, signals their potential baby-daddies' good health and good fatherhood. Iridescent plumage signals (accurately) to the peahens that the peacock is in good health. And if that feather formation compromises the survival of the peacock himself, it confers an advantage in mating, which ensures that that peacock reproduces, fulfilling that most basic aspect of biophilia, perpetuation of the genes.

THE LOGICAL FACT

So what else, aside from practicing a Buddhist tolerance for mystery, did our factophile do in the absence of data? He imagined future facts. He invented plausibilities.

Among the most legendary is Darwin's prediction of the existence of a moth with a twelve-inch tongue. Why was he even thinking such of such a creature? The rare Madagascan star orchid, a small white bloom, dangled a thin tube twelve inches long, storing a pool of nectar at its base. In 1862, Darwin hypothesized that some moth or another would have adapted to get at the nectary's treasure. Darwin's sometime rival Alfred Russel Wallace found this moth so plausible and likely, he asked a scientific illustrator to depict it, improbable proboscis and all. And sure enough, in 1903, scientists discovered what eventually proved to be a new species: a giant sphinx moth—"giant" meaning four inches long—with a tongue three times longer. This is the quintessential Darwin, imagining the truth before the arrival of the facts.

As noted earlier, a favorite pastime of Darwin's was, as he called it, "building castles in the air." In his notebooks, he mused on "the pleasure of the imagination, which [has] connection with poetry, abundance, fertility." For Darwin, as for a poet, the imagination is a field where a scientist's facts and mysteries may cross-fertilize. Once, he confessed his indulgence in a letter: "This confounded variation . . . is pleasant to me as a speculatist, though odious to me as a systematist." His love of speculation perhaps led him to chuckle in another letter: "A German writer was pleased to attribute the whole account [on barnacles] to my fertile imagination." The insult was a backhanded compliment on the prowess of Darwin's creativity. Darwin was ever sanguine in the face of insults—he once called himself a plodder, and he always plodded on. And even as he knew the German writer was wrong, he focused on how his critic "was pleased." Darwin's empathy had odd moments.

Darwin also speculated about the reasons for his own scientific success. He attributed it in his autobiography to a "fair share of invention." He felt that the naturalist should have "flexibility of mind," which he clearly exercised in creating his novel theories. He murmured in a notebook, "Jones said the great calculators, from the confined nature of their associations (is it not so in punning) are people of very limited intellects." By reductio ad absurdum, Darwin just called mathematicians dumb and jokesters smart. Intellect is inventive, not "confined."

Darwin's mind was unconfined. Legions of his speculations have been proven. As the Harvard biologist Ernst Mayr wrote, "It strikes me as almost miraculous that Darwin in 1859 came so close to what would be considered valid 125 years later." His imagination had clairvoyance. He was prescient. His castles in the air gave him a good view of wildernesses, where his mind wandered and wondered. Wondering was wandering into the new, and wandering led to wondering. He once remarked, "Perhaps one cause of the intense labour of original inventive thought is that none of the ideas are habitual, nor recalled obvious associations." Darwin embodies Einstein's point that "imagination is more important than knowledge."

In his books, he invited his readers to use their imaginations. "I must beg permission to give one or two imaginary illustrations," he wrote once, and, "It is good thus to try in imagination to give any one species an advantage over another." Free your mind, he challenged his readers: "What now are we to say to these several facts?" Thus he articulated his biophile's dictum:

Facts demand answers, so speculate. He urged his readers at diverse points to seek new points of view, to think like a bird, or like a "savage," or like a naturalist confronted by confusing evidence. Encouraging his readers to partner with him in analyzing, Darwin suggests, "we may suppose" and "let us imagine." Rhetorically, pedantically, he demonstrates that the imagination is a tool for thought experiments, not just for poets.

Along with imagination, logic can supply sense to facts, as he shows in *Descent* when airing the truth about sexual selection: "If we may assume that the females have the power of exerting a choice . . . all of the above facts become intelligible." Careful assuming, appeal to reason, the slant of the truth, guided by biophilia. His inventing was always grounded in the real, his imagination tempered by logic. As his 1883 eulogist L. C. Miall put it, "Darwin's power of reasoning from the seen to the unseen might be illustrated by nearly every chapter of his writings." Note that Darwin's castles in the air are *built*. He clearly "prefers to construct," in Fromm's terms. Darwin admired a long line of logic. Of a book on religious ideas about human evolution that he read as an undergraduate, he recalled: "The logic of this book . . . gave me as much delight as did Euclid. . . . I did not at that time trouble myself about Paley's premises; and taking these on trust I was charmed and convinced by the long line of argumentation." The modern philosopher W. V. O. Quine has pointed out, "If sheer logic is not conclusive, what is?" Darwin felt he wrote *Origin* with rigorous thought: "Some of my critics have said, 'Oh, he is a good observer but has no power of reasoning.' I do not think this can be true, for the *Origin of Species* is one long argument from the beginning to the end, and has convinced not a few able men." Darwin's imagination as both creative and logical reflects a fruitful synthesis of two components of Fromm's biophilia: the compulsion for the new and a reliance on reason.

OTHER THAN FACTS

Darwin used other inner technē to see and build: intuition, emotion, and a sense of the limits of the rational. Is this surprising for a factophile? Only someone who knows what facts can do knows what they cannot. Only someone who loves reason can see that it may be necessary but not sufficient. As noted earlier (see chapter 1), reason cannot account for Nature's paradoxes.

Darwin used intuition to build his theories. He possessed the kind of intuition described by the English philosopher Bertrand Russell when presenting the ideas of the French philosopher Henri Bergson:

> In man as he now exists, intuition is the fringe or penumbra of intellect; it has been thrust out of the centre by being less useful in action than intellect, but it has deeper uses which make it desirable to bring it back into greater prominence. Bergson wishes to make intellect "turn inwards on itself, and awaken the potentialities of intuition which still slumber within it." . . . Intuition is . . . synthetic rather than analytic. It apprehends a multiplicity, but a multiplicity of interpenetrating processes, not of spatially external bodies. There are

> in truth no things: "things and states are only views, taken by our mind, of becoming. There are no things, there are only actions." This view of the world, which appears difficult and unnatural to intellect, is easy and natural to intuition. . . . Intuition alone can understand this mingling of past and future.

Oddly enough, this passage also describes evolution perfectly: "things and states are only views, taken by our mind, of becoming" might be an alternate description of species, which are temporary. Over the long haul of time, they are continually becoming, evolving. Darwin sensed this, that every life form is a temporary adaptation to circumstances, a suspended process, a momentary stasis. And Bergson's phrase "intuition . . . is synthetic rather than analytic" describes Darwin perfectly. He could analyze with the best of them, of course, but when he built his castles in the air and wrote books of pure theory, he was synthesizing, seeing, as Fromm wrote, "the whole rather than only the parts." Likewise Darwin saw that every life form is *a mingling of past and future*—inheritance and adaptations from its past and the inevitability of change in its future.

There is no word in the English language that means having correct intuitions all the time and imagining future realities, though "savvy" and "prescient" approach these ideas, respectively. But some people have a great "capacity for union with that which is to be known," as the modern biographer Evelyn Fox Keller wrote when describing the botanist and Nobel laureate Barbara McClintock. Remember Darwin's grandfather Erasmus, who imagined airplanes. To know, often, with no proof of what will be found true and what will be made real—this is the quantum leap, the intellectual intuition, that distinctive reality-based imagination. In a funny example: Darwin imagined dyed flowers. He wrote in his autobiography that as a child, "I told another little boy . . . that I could produce variously coloured Polyanthuses and Primroses by watering them with certain colored fluids, which was of course a monstrous fable, and had never been tried by me." Dyed flowers may be monstrous, but today they are at the corner florist. Darwin suspected that this "fable" lingered in his conscience because he had lied, but perhaps it had, rather, loitered, awaiting "the outbreak of the definite." His imagination traveled from the nonexistent to the eventual, from the potential to the actual, from present data to methods two centuries away, from the crackpot to the commonplace. I like the word "factspeculating." Seeing the facts-to-be.

As part of his intuition, Darwin's feelings could reflect unknown facts in eerie ways. For example, he raved about the Brazilian rainforest for days in his journals and in expansive paragraphs in his books. (One such passage appears in chapter 6.) He resonated intensely with the unique beauty of the region, beauty above every other locale he admired during that five-year trip around the world. More than a hundred years later, we have learned that this region contains an abundance of life forms unmatched elsewhere on Earth. He had resonated to a fact science did not yet know. And his emotion serves as an example of another insight noted earlier, again offered by Fox Keller about McClintock: "Good science can not proceed without a deep emotional investment on the part of the scientist." Here, Darwin "knew" a fact through love.

Keller continues, "McClintock's feeling for the organism is . . . a longing to embrace the world in its very being through reason and beyond." This is another good definition of biophilia. Keller's "deep reverence for nature" shows "a different image of science from that of a purely rational enterprise," one that has a long history in natural philosophy. A passion for nature is, of course, traditional outside of science, serving, for example, as a key tenet of the Romantic movement in Western culture, as well as the basis for many Indigenous worship systems. The rejection of the rational has its tradition, with the philosopher George Santayana stating, "The ideal of rationality is itself as arbitrary, as much dependent on the needs of a finite organization as any other ideal." If Darwin had waited for evolutionary theory to seem completely rational, he might not have published. And in a passage that might serve as Darwin's manifesto, Keller wrote:

> For McClintock reason—at least in the conventional sense of the word—is not by itself adequate to describe the vast complexity—even mystery—of living forms. Organisms have a life and order of their own that scientists can only partially fathom. No models we invent can begin to do full justice to the prodigious capacity of organisms to devise means for guaranteeing their own survival. On the contrary, "anything you can think of you will find." In comparison with the ingenuity of nature, our scientific intelligence seems pallid.

THE VILLAINOUS FALSE FACT

In his respect for truth, Darwin was scrupulous about what "we may infer" and what "many facts clearly show" and in referring for support to "details. . . . I have collected and elsewhere published." He carefully announces when he is, instead, "guided by theoretical consideration" and when "Dr. Hooker permits me to add that he thinks the following statements are fairly well established." Every fact has a provenance for Darwin, even his own, which he specified. "I have endeavored to test this numerically . . . and as far as my imperfect results go, they confirm this view. I have also consulted some sagacious and experienced observers, and . . . they concur." He treated facthood with care.

He knew that facts are constituted through research and thought. For Darwin, the definition of "fact" was the simplest, most Latinate, dating back to sixteenth-century England. A fact was "something made or done," which seems like two plus two. Yet there was and still is another common dictionary definition of fact as "a statement that can be proven true or false." This one would have been nonsense to Darwin, who might have asked, "Proven how? Do we have proof of the efficacy of the method of proof?" This is a cruel mockery given the existence of the "false fact," which he felt to be a dangerous species of fact. The love of truth in biophilia entails an abhorrence of the false.

Jaded by postmodernism and Twitter and today's scourge of "alternative facts," we may forget that the false fact is not new but existed in the lexicon in Darwin's day. The verity of any particular datum has long been a slippery business. Darwin felt obliged to warn his readers about false facts in the closing passages of *Descent*: "False facts are highly injurious to the process of science, for they often long endure; but false views, if supported by some evidence, do little harm, as every one takes a salutary pleasure in proving their falseness; and when this is done, one path towards error is closed and the road to truth is often at the same time opened."

He had, then, a sense of the villainy of false facts. There were several fundamental putative facts current in his day that were intellectual impediments to scientists, such as the physics "proving" that the solar system was too young for evolution to have occurred. Lord Kelvin's calculations of the age of the sun certainly rated as false facts. Darwin wrote: "I feel a conviction that the world will be found rather older than Thomson [Kelvin] makes it," he wrote. But knowledge of physics at the time was insufficient to support Sir George Lyell's hypothesis about the longevity of the Earth, a time span that aligned with evolutionary theory. Scientists discredited his theory because he lacked facts, pleasing those who believed the biblical story of Creation. But Darwin trusted the logic and the geology of his good friend and colleague. As ever, his intuitive, no-proof-required sense of the real was correct.

Darwin was able to remain uncertain in the face of a lack of facts or consistency, in that Zen-ish biophile way of being able to hold two contradictory thoughts in mind at once (see chapter 1), being able to accept mysteries without needing to resolve dichotomies, embodying Keats's ideal. After an exposition assessing his evidence in *Origin*, Darwin still writes that he "can form no decided opinion" on "a laborious collection of all known facts," which is to say that lots of facts can still be insufficient for determining opinion, let alone truth. As noted earlier, he could accept ignorance of Nature's genius. "Science has not yet proved the truth of this belief, whatever the future may reveal," he remarked once. But he had faith that the truth will out, via the long labors of science—and logic and imagination and intuition. He had the sage's patience with knowledge "unfolding," Fromm's biophilia yet again.

THE TOUCH OF THE REAL

What Darwin saw often struck him as having self-evident import: "Any one who is not convinced by such facts as these, and by what he may observe with his own dogs, that animals can reason, would not be convinced by anything that I could add," he wrote in *Descent*. Maybe I imagine exasperation in that sentence, and possibly the exasperation is because the subject is dogs. But when a person grasps a fact intuitively, that person experiences a paradox: knowledge without data to sanctify the knowledge. And the intuitor may feel a consequent frustration in realizing that everyone else does not "know" this unproven fact. Worse may be the futility of finding and giving facts. A biophile may experience culture shock now and again. Nevertheless, Darwin collected and created and promoted facts all his life. As he aged, his life was all about facts and more facts, truth and more truth. He reflected, "My mind seems to have become a kind of machine for grinding general laws out of large collections of facts." This was a labor of love—and a labor of the love of life.

Chapter Four

THE DANCE OF PLANTS, THE ROOTS OF MIND

Every little pine-needle expanded and swelled with sympathy and befriended me.

—Henry David Thoreau

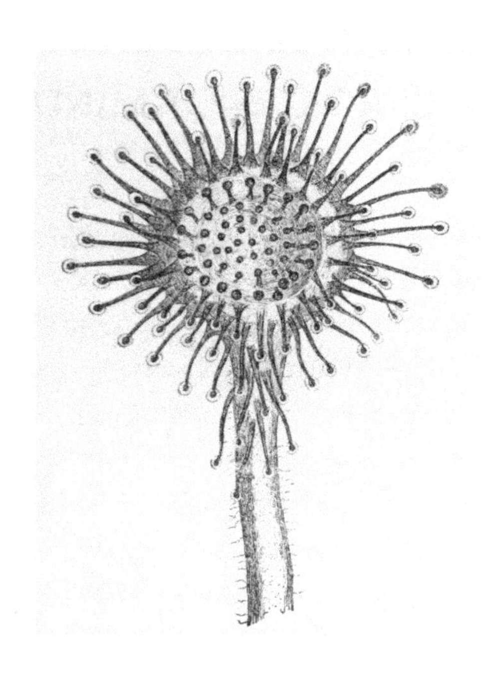

FIGURE 4.1 Darwin's hothouse carnivore: The insectivorous *Drosera.* Carnivorous plants had their moment in the sun with the twentieth-century classic *Little Shop of Horrors*. But Darwin became fascinated by *Drosera rotundifolia*, a cousin of the Venus flytrap, when he first pondered it in 1860. Colloquially the common sundew, *D. rotundifolia* is a gaudy herb of green, red, and gold, sparkling with droplets of a scented glue that lures and ensnares insects. Purple fluid courses through its cells, and its flower is hermaphroditic. *D. rotundifolia* curdles milk, eases whooping cough, withers warts, and regenerates in inhospitable locales. Scientists today are exploring its potential for tissue engineering and self-repair. It is no surprise this powerful creature ensnared Darwin. Fifteen years after plucking *Drosera* from a Sussex heath and tempting it with everything from meat both raw and roasted to gold leaf, cork, and dried grass, Darwin declared in *Insectivorous Plants* that not only did the plant sense what was and was not food, but "Drosera possessed the power of dissolving solid animal matter . . . and . . . the glands absorb the digested matter. . . . No such power was before distinctly known to exist in the vegetable kingdom." Many of his colleagues could not believe in the possibility of carnivorous plants and the discretionary powers such ingestion requires. Darwin wrote, in caveat, that "no one supposes the Sensitive-plant is conscious," but his forte was supposing what "no one supposes," and *Drosera* inspired him to private musings on plant consciousness.

Source: Reproduced with permission from John van Wyhe, ed., *The Complete Work of Charles Darwin Online*, 2002–, http://darwin-online.org.uk/.

DARWIN TICKLED the innards of a flower, impersonating a bird. He confessed this to the botanist Sir Joseph Hooker. Darwin fooled the blossom, he crowed in a letter, advising, "It is worth-while to pass your little finger like a bird's beak as if to enter within the little petal at the base for nectar, & see how neatly the two other oddly shaped blue petals open and expose pollen." He saw a mechanism of survival in "how neatly the . . . petals open," an action he considered informed and deliberate. Of another plant, Darwin wrote that he wished to "teach it to close by itself" by "covering it up daily for half an hour." He schemed to trick it to override its chronobiology. That is to say, he thought it might learn to adapt. Along with Hooker, Darwin's wife, Emma, was also privy to what he was up to out back: "He is treating *Drosera* [which eats insects] just like a living creature, and I supposed he hopes to end in proving it to be an animal." Indeed, he admitted in a letter, "It has always pleased me to exalt plants in the scale of organized beings." And to tickle them.

Of the insectivorous *Drosera*, Darwin wrote: "If any bit of dry moss, peat, or other rubbish, is blown on to the disc [the center of the flower], as often happens, the tentacles clasp it in a useless manner. They soon, however, discover their mistake and release such innutritious objects." Note that the tentacles note changes in their environment and make discoveries, as scientists do: they assess these against certain criteria, then act on judgment that is correct. A scientist, however, may monkey with the process now and then. Only once in his notebooks did Darwin remark that plants were "without thought." The preponderance of his notes and published work explores their "mental powers." He was sure, for example, that plants have memories.

The fact plain to Darwin was that a plant tip curling away from a drop of acid resembles a toddler hiding behind a trusty companion upon seeing a stranger. Both indicate a life-loving wariness in the face of the new. The plant is correct in recognizing the danger of the acid. Darwin saw plants as acting in ways that ensure their lives. Language failed Darwin when plants thus proved themselves smart and mysterious, just as it had failed Rilke when he felt connected with the heart of a dog. Like Rilke, Darwin felt compelled to invent a verb: "circumnutate," to describe how a tendril tip moves when reaching for the life-giving. The stem of a seedling, he wrote, "bends successively to all points of the compass." Plants seek to grow, he discovered, with an endless slow-motion twirl, responding to light, air currents, water, contact, gravity, the sun, and Earth's spin. He described the tip of a climbing plant thus: "The whole terminal portion exhibits a singular habit, which in an animal would be called an instinct; for it continually searches for any little crevice or hold into which to insert itself." This plant is *searching*, that is, proactive; to Darwin's eye, the plant has a purpose.

Darwin believed, as noted earlier, that "thought originated in sensation," in keeping with the eighteenth-century philosopher Edmund Burke's notion that "the senses are the great originals of all our ideas." So he saw sensate plants exhibiting their mental powers. His biophilia resonated to and thus discovered the plant's powerful and empowered urge to survive. For Darwin, the seedling's dance was choreographed by purpose. The circumnutating tendril tips reflected the one living spirit, basic biophilia. An exemplum of his own "one thinking principle." Looking at plants, he sensed the autopoietic dictum of Maturana and Varela (see chapter 1): "living systems are cognitive systems, and living as a process is a process of cognition." Plants nurtured Darwin's thought as he sought the roots of mind in the simpler organisms.

MINDING THE PLANT

Darwin put his finger on thought as a puzzling epiphenomenon when he scribbled in a note he consigned to the "Old & Useless Notes" folder: "The reason why thought etc should imply the existence of something in addition to matter is because our knowledge of matter is quite insufficient to account for the phenomena of thought." This scrap also shows how he is ever aware of the unknown (see chapter 3), an intellectual stance helpful when searching for thought in plants. And his work shows how biophilia may be understood as primal thought.

To return to the carnivorous *Drosera* in Darwin's little shop of horrors and exotica: Darwin scrutinized his little green friends for rational behavior, asking himself such questions as: "Have plants any notion of cause & effect; they have habitual action which depends on such confidence." Eventually, he decided that *Drosera* did indeed think, which may have shocked the world but probably was no surprise to Emma. He declared in one letter, "I sometimes think *Drosera* is a disguised animal," and in another, he raved about this "most sagacious animal." Even as he wrote tongue in cheek, he wrote what his biophilia felt.

Darwin noted that "movement of sensitive plants . . . shows a local will, though perhaps not conscious sensation." When he juxtaposes "local will" against "conscious sensation," he contradicts his own sense that thought emerges from sensation; local will and conscious sensation are not, in fact, opposite for him but happen in concert, in Möbius fashion. "Local will" implies effort, a choice in accord with a drive to survive, a choice to act a certain way predicated on the reasoning of staying alive. It implies consciousness of sensation *and* a concomitant reaction that is in fact an act of will—as Darwin wrote, "It is easy to conceive such movements & choice," and he went on to ask himself, "How does consciousness commence; where other senses come into play." A survival mechanism in response to a sensation is clearly in the realm of thought, in Darwin's view, consciousness commencing when the senses come into play. Darwin used the word "brain" time and again in *The Power of Movement in Plants*, as when he wrote that the "tip of the radicle [root] . . . acts like the brain of one of the lower animals." This was more than just a metaphor—Darwin was seeking to root the origin of the mind and the phenomenon of consciousness in the simpler organisms.

So for Darwin, the self-preservation, the survival mechanisms, that is, the plant's primal biophilia, hinted at a brain generating behavior that accords with a struggle for existence. He found many traces of biophilia but no brain. All that he could muster was that his insectivores possessed "matter at least in some degree analogous in constitution and function to nervous matter." He was well aware, of course, that his failure to discover a brain could be explained by a lack of technē (see chapter 3), not necessarily caused by the lack of a brain.

Surely Darwin would have been thrilled—and perhaps not surprised—to learn what today's botanists know: that plants recognize their kin, emit chemical warning signals about oncoming threats, and engage in other exchanges for the sake of survival, exchanges that can be called communication, even mutual aid, one of the hallmarks of *Homo sapiens* and other social species (see chapter 6). No one knows today where the dividing line is between sensation and sense, feeling and thought, neither in plants nor in humans. But if Darwin could not find the physical traces of a mind, his work still shows that thought is rooted in survival, aka the original biophilia.

THE EMBODIED MIND

Seeking elsewhere in the spectrum of life to trace the visceral origins of *Homo sapiens*' reasoning mind, Darwin praised the ant. He was certain that there could "be extraordinary mental activity with an extremely small absolute mass of nervous matter. . . . Under this latter point of view, the brain of an ant is one of the most marvelous atoms of matter in the world, perhaps more marvelous than the brain of man." Here, as ever, we find the man inclined to jab at *Homo sapiens*' self-esteem. Also worth noting is that the ant that inspires Darwin's praise is the subject of a book coauthored by Edward O. Wilson, whose description of his own biophilia describes Darwin's as well (see chapter 1). Biophilia and evolution share the concept of oneness, and in searching for thought in every corner of the biological world, he was looking for yet one more oneness, more evidence for our common continuity, the one progenitor of all living forms.

Darwin was sure there was nothing mystical about the emergence of mind, even if, as he said, our knowledge of matter was not sufficient to explain the mind. But he knew at least the brain was matter, organic, biological matter, a nakedly, scandalously physical organ, evolved along with the rest of the material body in response to the primal biophilia. The material mind accorded with his ideas on how thought begins in sensation: he told himself that he would be amazed "if mind of animal was not closely allied to that of men, when the five senses were the same." This is, after all, the man who, as already noted, once groused in a notebook, "Having proved men's and brutes' bodies on one type, almost superfluous to consider minds." To Darwin, the human's sacrosanct mind was little more than a "thought-secreting organ," a tool developed by natural selection for survival. He scrawled this kind of phrase in notebooks and book margins. He did not use it in public or in his books.

He returned to this theme time and again. "Thought . . . obeys the same laws as other parts of structure," a view tantamount to heresy for many. And likewise he scribbled secretly: "Why is thought being a secretion of brain, more wonderful than gravity a property of matter. It is our arrogance, our admiration of ourselves." Thought to Darwin was no more sacred than the acid of the stomach, a biological function driven by input and reflex. He told himself firmly: the "problem of the mind cannot be solved by attacking the citadel itself.—The mind is function of body." What he showed is mind emerging from biology, via survival, that is, via biophilia. He prefigures the modern neuroscientist Antonio Damasio, who tells us that "the brain is the body's captive audience."

Darwin began with the body when he began to explain the evolution of *Homo sapiens* in *The Descent of Man*. One of his first key points was to illustrate "homologous structures" across mammals. As shown earlier, he paired drawings of the backbones of an embryonic human and an embryonic dog—the dog serving more congenially than the arboreal quadruped with prehensile feet that he would introduce later in the book. But he was forthcoming with his book's unwelcome news: "It is notorious that man is constructed on the same general type or model with other mammals. All the bones in his skeleton can be compared with corresponding bones in a monkey, bat, or seal. So it is with his muscles, nerves, blood vessels, and internal viscera. The brain, the most important of all organs, follows the same law, as shewn by Huxley and other anatomists." Huxley, according to Darwin's son Francis, even said that a plant is "an animal enclosed in a wooden box." If Darwin had not quite found that animal, had not quite traced the brain to the plant, he could at least link it to the monkey, the bat, and the seal.

HOW TO CHANGE YOUR MIND

And indeed, in looking at social animals such as monkeys, bats, and seals, Darwin examined reason more deeply. In a notebook scrawl that became scientific dogma, he declared: "I say grant reason to any animal with social & sexual instinct[s]." Animals reasoned, he was sure. Elsewhere in a notebook, Darwin declared, "All Science is reason acting, systematizing, on principles, which even animals . . . know." In the same vein he noted, "The more intellectual emotions and faculties . . . form . . . the basis for the development of the higher mental powers."

While Darwin does say "higher" mental powers, he is ambivalent about the highest: reason (see chapter 3). He will pay tribute to reason, telling us it "stands at the summit" of the human mind. He will acknowledge that "of the high importance of the intellectual faculties there can be no doubt, for man mainly owes to them his preeminent position in the world." Then he starts the *ifs*, *buts*, *yets*, *then agains*, and *maybes*. We share many of those "intellectual faculties" with other animals, he pointed out. For every compliment on the human brain's "extreme development" he reminds us that "man is descended from some lower form" and that "there is no fundamental difference between man and the higher mammals in their mental functions."

Darwin pummeled the mind of "Man" from every angle. He replayed the theme that reason and instinct are "often difficult to distinguish." Like those mushing dogs who halt at the sight of thin ice—intelligence or instinct? But at least once, he jotted down his own clear answer: "People often talk of the wonderful event of intellectual Man appearing.—the appearance of insects with other senses is more wonderful." He exalted the insect, as well as the plant, and demoted *Homo sapiens sapiens*, formally named twice for wisdom. This was one of the more traitorous tasks a Victorian scientist might undertake. But Darwin forged ahead and lauded reason in the birds and the bees. And he was crystal clear at least once in *Descent*: "The mental faculties of man and the lower animals do not differ in kind, though immensely in degree." Darwin's biophilic instincts about the brain became foundational to modern cognitive and evolutionary psychology: Within two decades, the legendary William James began arguing for an understanding of "the embodied mind"—that very mind being a bodily organ shared with animals.

Along with knocking the vaunted human mind off of its pedestal, Darwin traced continuities of thought across species, erasing differences, finding equals where hierarchies had stood. Some of the higher mental functions he finds across many species are dreaming, imagination, memory, responsiveness to emotion, and reason. As noted earlier, when Darwin examined dreaming, we find the dog prompting an early insight: "A Dog whilst dreaming, growling yelping & twitching paws . . . shows their power of imagination . . . think well over this;—it shows similarity of mind." Given the premium Darwin placed on the imagination (see chapter 3) as a key activity of mind, this "higher faculty" in dogs is telling for Darwin.

Along with arguing for continuities across species in matters such as reason and dreaming, Darwin also traced continuities in emotion. In *The Expression of the Emotions in Man and Animals*, written in 1872, his primary goal, wrote one scholar, was to "substantiate evolutionary theory through the universality of certain emotions—those growing from the basic instincts of survival." He had been thinking about emotions for more than thirty years; in 1838, he observed in a notebook, "Those emotions which are strongest in man, are common to other animals & therefore to progenitor far back." In 1854, he sought to define his terms: "What is emotion?" The question is both simple and radical, posed in his let's-start-from-scratch inimitable way.

He tasked himself in *Expression* with portraying the inner life of a large range of species and demonstrating the commonalities. This inner life can only be inferred by outer expressions, of course, so he catalogued facial, gestural, postural, and vocal behaviors that indicate emotion. Across life forms he found these "ancient expressions": the needy whine, the warning snarl, the "scream of agony," the sigh of "meditative tranquility," and more. Darwin found that performing one or another of these ancient expressions were the baboon, bird, cow, elephant, horse, human, hymenoptera, insect, monkey, ourang outang, parrot, snail, spider, and turkey. Pressing in on his point, Darwin wrote, "It is not a little remarkable that those sounds which are involuntary are common to animals." It was a subtle way of restating what he had jotted down in his notebook regarding our common "progenitor."

But eventually, as previously quoted, he straightforwardly stated it: "Man . . . uses, in common with the lower animals, inarticulate cries to express his meaning, aided by gestures and the movements of the muscles of the face. This especially holds good with the more simple and vivid feelings, which are but little connected with our higher intelligence. Our cries of pain, fear, surprise, anger . . . the murmur of a mother to her beloved child, are more expressive than any words." Note that Darwin adds a characteristic grace note, bringing in a loving mother to aid his argument. The modern literary scholar Stanley Edgar

Hyman wrote of *Expression*: "The final sense we get is of a community of feeling and reaction in infant and adult, elephant and keeper, degraded woman and galvanized man, Darwin and a small monkey sharing his snuff."

If continuity of feeling spoke of common ancestry, it also spoke of the common need to survive, aka biophilia. And here is where Darwin's denigrations of reason and admiration of emotion converge: "The mind of man is no more perfect, than the instincts of animals," he admonished. He noted: "instincts is [*sic*] a modification of bodily structure . . . & intellect is a modification of . . . instinct." Mind was never the sine qua non for Darwin, even as he sought to explain it.

A NOTE FOR THE ANNALS OF ANTHROPOMORPHISM

In studying emotional commonalities, Darwin used "his usual process of empathy and identification," as Hyman described it, and manifested an "affiliation with life," to return to Wilson's definition. Darwin felt organized beings' inner lives, their drives, and the logic of their actions, whether those organized beings were human or otherwise. His granddaughter noted that "his sympathetic participation in the lives of the creatures he observed helped him to understand their habits." So it happens that he exhorted us: "With respect to female birds feeling a preference for particular males we must bear in mind that we can judge of choice being exerted, only by placing ourselves in the same position." Say what? Walk a mile in the shoes of a female bird deciding with whom to mate? Think what a bird thinks? What kind of person could think this is not merely a wild flight of fancy? Someone who often takes such leaps (see chapter 3).

Darwin wrote anthropomorphically because his theory pushed him to think anthropomorphically. In a complement to the rigor of his taxonomic work and as a testament to the diversity of his modes of inquiry, he conducted empathetic thought experiments, for example, when he speaks for the thoughts of a female bird or of a dog, when he imagines the reasons for the actions of *Drosera*, a bee, a chimpanzee, and so many others. He commented in his travelogue that one night in South America he could not eat his dinner because he could hear his horse gnawing at its post and knew it was hungry.

In Darwin's day, natural philosophers and scientists stigmatized anthropomorphism. Darwin's exposed him to much criticism and contributed to his disrepute. But, again, yesterday's discreditable anthropomorphism is today's anecdotal cognitivism. Darwin knew in his bones that we share feelings with animals. He once emphasized in a notebook: "Animals—whom we have made our slaves we do not like to consider our equals.—Do not slave-holders wish to make the black man other kind?—Animals with affections, imitation, fear of death, pain, sorrow for the dead—respect."

For Darwin, animals were brethren, as in: "If we choose to let conjecture run wild, then animals, our fellow brethren in pain, disease, death & suffering; our slaves in the most laborious work, our companion in our amusements . . . from our origin in one common ancestor, we may all be netted together."

The modern primatologist Frans de Waal, who studies culture in animals, comments that "anthropomorphism acknowledges continuity between humans and animals" and is "part and parcel of the way the human mind works." This aspect of *Homo sapiens* may have been complete heresy in Darwin's day, but it was A-B-C to de Waal and a quintessential aspect of the biophilia that Wilson describes as being at one with all of life.

So in the context of biophilia, Darwin's anthropomorphism accords his sense of commonality, of the one living spirit. The way he wrote *The Descent of Man*, as the Darwin scholar Will Durant has noted, was to show that "human thoughts and actions were explained in terms of animals instinct, and . . . animal behavior . . . in terms of human thoughts and feelings." This is true of Darwin's style and thought, of course, but it is also true that often he did not distinguish between human and animal, between instinct and thought, between a binary and a Möbius. As another modern Darwin scholar, David Kohn, wrote: "Thus was anthropomorphic zoology combined with zoomorphic anthropology in effecting the unification of animals and man, matter and mind, nature and morality. Darwin's anthropomorphism was the corollary of his rejection of anthropocentrism, and this rejection in turn followed from his meta-theoretical commitment to the principle of continuity in nature." And of course, said "meta-theoretical commitment to the principle of continuity in nature" might be paraphrased as Darwin's profound sense of the one living spirit, that singular case of biophilia.

But even as Darwin spoke for so many species and for oneness, he also knew that each species has its sapience, one suited to that species' survival, a one-of-a-kind adaptation evolved through biophilia to live and make life. Human sapience suits us but is no better than the ant's. Darwin respected each species' sapience as a "truth of one," as Gandhi might say. And what of the sapience of the flower he tickled? He tricked the flower into opening its petals and revealing biophilia, his "one thinking principle."

Chapter Five

THE VARIETIES OF PASSIONATE EXPERIENCE

Only connect!

—E. M. Forster

FIGURE 5.1 A feline Mona Lisa.

Captioned by Darwin "Cat in an affectionate frame of mind," this illustration would likely not grace today's scientific pages. But in Darwin's 1872 *The Expression of the Emotions in Man and Animals*, the leg-rubbing cat by Thomas William Wood displayed Darwin's hypothesized "principle of antithesis." The cat's body language here is the opposite of a cat's angry or threatening posture, which features flattened ears, bared teeth and claws, erect hair, and a taut crouch or dramatic arch to "appear terrible," he wrote. Expressions of pleasure—in behaviors, sounds, signals, and body language—drew Darwin's interest in particular, and he wrote that the cat's "purr of satisfaction . . . is one of the most curious." He noted more generally of expressions that "the power of intercommunication is certainly of high service to many animals." The modern psychologist Paul Ekman has discussed the smile as according with Darwin's principle of antithesis. It is the clearest of human signals, which is interesting inasmuch as biophilia inheres in the pleasures a smile implies. Darwin was scrupulous about his images and likely meant this cat to display this odd half-smile—not quite the Mona Lisa or the Cheshire Cat but anthropomorphic, human, tantalizing. The images in *Expression* contributed to its success on the market, and the book was yet another bestseller for Darwin.

Source: Reproduced with permission from John van Wyhe, ed., *The Complete Work of Charles Darwin Online*, 2002–, http://darwin-online.org.uk/.

AS DARWIN TRAVELED through South America, he was wowed by the local shepherd dogs, as he told readers of *The Voyage of the* Beagle:

> While riding, it is a common thing to meet a large flock of sheep guarded by one or two dogs, at the distance of some miles from any house or man. I often wondered at how so firm a friendship had been established. The method of education consists in separating the puppy, while very young, from the bitch, and in accustoming it to its future companions. An ewe is held three or four times a day for the little thing to suck, and a nest of wool is made for it in the sheep-pen; at no time is it allowed to associate with other dogs, or with the children of the family. The puppy is, moreover, generally castrated; so that, when grown up, it can scarcely have any feelings in common with the rest of its kind. From this education it has no wish to leave the flock, and just as another dog will defend its master, man, so will these the sheep. It is amusing to observe, when approaching a flock, how the dog immediately advances barking, and the sheep all close in his rear. . . . A pack of the hungry wild dogs will scarcely ever venture to attack a flock guarded by . . . these faithful shepherds. The whole account appears to me a curious instance of the pliability of the affections in the dog. . . . The shepherd dog ranks the sheep as its fellow-brethren, and . . . the wild dogs, though knowing that the individual sheep are not dogs, but are good to eat, yet partly consent to this view when seeing them in a flock with a shepherd-dog at their head.

In high-gear anthropomorphism, Darwin articulates how a dog *ranks*, *knows*, *has no wish*, *consents*, and responds in calibration to other dogs' ideas. As his theoretical work progressed, this psycho-ethological portrait became a picture not only of the dog's pliant nature, but of pack loyalty, of the social organism's instinct to affiliate with, care for, and bond with.

Darwin painted many lush scenes of interspecies love. Early in *The Descent of Man*, he extolled the "capacious" heart of a captive baboon who adopted puppies, kittens, and young monkeys; she groomed all in happy oblivion to morphological variation. He also offered his readers an anecdote about a dog he knew "who never passed a great friend of his, a cat, without giving her a few licks of his tongue, the surest sign of kind feeling in a dog." And he lionized a "sympathetic and heroic" small monkey who attacked a baboon attacking a zookeeper whom the small monkey liked. As with his description of the South American shepherd dogs, he gives verve to his "amusing" vignettes, beguiling with the linguistic charm that contributed to the commercial success of *Voyage*.

Charm notwithstanding, Darwin delved deep into family bonds as a topic of import regarding instinct. He studied, in fact, a range of instinctive emotions: parental love, sexual love, and joy. Understanding the varieties of passionate experience—their various means and goals and biological history—was part of Darwin's evolutionary project. And it was, in fact, tantamount to setting out to prove what the philosopher Edmund Burke had said the century before, that "the passions are the organs of the mind." In characterizing these three powerful passions, Darwin focused on biophilia's life-affirming good—whether that good is moral, physical, essential, or psychological. He chronicles kindness and sexual pleasure, affiliation and its antithesis aggression, altruism and the joy of a child. The theme of his insights might be summed up as: we take joy in living, in loving, and in being joyful. His work accords with Fromm's definition of the biophile: "The person who fully loves life is attracted by the process of life. . . . Biophilic ethics have their own principle of good and evil. Good is all that serves life. . . . Good is reverence for life." Darwin's thoughts about passion—that passion is good because it is life-giving—are another exercise in his biophilia finding biophilia.

LOVE THE ONE YOU'RE WITH

Family values were personally sacred for Darwin. In *Voyage*, he recounted the story of a South American colonialist who quelled a slave revolt by threatening to sell all the women and children away from their families. He was appalled about "those atrocious acts, which can only take place in a slave country."

As with so many phenomena Darwin noticed, the reflex to bond has been confirmed as a crucial instinct hardwired in social beings. The identification of this innate instinct resulted from legendary experiments led by the Nobel laureate Konrad Lorenz; he termed it "imprinting." The jackdaws who mistook his feet for their mother proved that the jackdaw's heart and mind would label *mother* anything they saw at just the right moment. One implication is that affiliative love is based on the need for survival. Such love is a lingua franca of every species because it is a facet of the most biological biophilia, the dire drive of life to live. Young creatures really do run around asking, "Are you my mother?" as in the children's classic, *Are You My Mother?*

Moving to the parental side of this dynamic, Darwin wrote in *Descent*, "One of the strongest of all instincts [is] the love of . . . young offspring." He summoned up an instinctive parental urge, which seems to him to be a reflexive kindness, a "benevolence," one of his favorite words. This urge would be fundamental to a biophilia that encompasses selfish-gene theory, since parental altruism serves the self's survival inasmuch as it preserves and perpetuates the self's genes. In accord with Fromm's assertion that "Good is all that serves life," Darwin's idea of benevolence encompasses all of altruism, morality, and connection. Noted earlier was this famous speculation: "Moral sense may arise from strong instinctive sexual, parental, & social instinct, giving rise [to] 'do unto others as yourself.'" This musing emerged formally as:

> Looking at Man, as a Naturalist would at any other mammiferous animal, it may be concluded that he has parental, conjugal, and social instincts. . . . These instincts consist of a feeling of love and sympathy, or benevolence. . . . We see in other animals they consist in such active sympathy that the individual forgets itself & aids & defends & acts for others at its own expense. . . . Moreover any action in accordance to an instinct give[s] great pleasure. . . . Therefore in man we should expect that acts of benevolence towards fellow creatures, or of kindness to wife and children would give him pleasure, without any regard to his own interest.

So altruism is not confined to *Homo sapiens*. Darwin argued that our ancestors

> must have acquired the same instinctive feelings which impel other animals to live in a body; and they no doubt exhibited the same general disposition. They would have felt uneasy when separated from their comrades, for whom they would have felt some degree of love; they would have warned each other of danger, and have given mutual aid in attack or defence. All this implies some degree of sympathy, fidelity, and courage.

He made the very same point regarding dogs, who, he wrote, "have long been accepted as the very type of fidelity and obedience."

Social packs enhance survival. Biophilia may indeed begin as self-preservation and diversify into more complex modes for pursuing life, as described in chapter 1, and Darwin saw just such a pattern in the emergence of domestication, writing, "Animals in the first place [were] rendered social . . . in the same manner as the sense of hunger and the pleasure of eating were, no doubt, first acquired in order to induce animals to eat." Eating and bonding alike served survival. Social order operationalizes biophilia.

Darwin repeated it and repeated it. He asserted in *Descent* that "sympathy . . . is of high importance to all those animals which aid and defend each other. . . . Those communities, which included the greatest number of the most sympathetic members, would flourish best and rear the greatest number of offspring."

With this theory, Darwin endorses not survival of the "red in tooth and claw"—as Alfred Lord Tennyson described the survival of the fittest—but survival of the kindest, of the "most sympathetic," of the ones with stronger biophilia.

The crucial sympathy abides among even tigers and lions, he reminded his readers, figuring even the ferocious as altruistic, dramatically throwing a little paradox into the mix. In *Descent*, he added, "each man would soon learn that if he aided his fellow-man, he would commonly receive aid in return. From this low motive he might acquire the habit of aiding his fellows; and the habit of performing benevolent actions certainly strengthens the feeling of sympathy which gives the first impulse to benevolent actions." Moreover, he believed that as the human animal became more intelligent, it became more benevolent, in what we call today a virtuous cycle. He wrote, "We may expect that in future generations . . . virtue will be triumphant."

And it is not just the benevolence per se that enacts biophilia but that the benevolent are responding to their own instinct to aid, which also feels good. Darwin made the point again and again: "The impulse which leads certain animals to associate together and to aid each other [is] impelled by the same sense of satisfaction or pleasure which they experience in performing other instinctive actions." He had a dog example at the ready to support his views: "dogs take pleasure, when doing what they consider their duty. . . . They feel pleasure in obeying their instincts." Darwin also offered the examples of the "peasant" and "philosopher" whose "thoughts are most pleasant when the conscience" tells the mind "good has been done." The impulse to aid others and the reflexive feeling good in consequence is

Darwin again describing Fromm's biophilia. All of Darwin's benevolent instincts, "sexual, parental & social" are pro-life, biophilia manifest. Altruism is the default state in Darwin and Fromm.

Darwin's instinctively moral creature is, then, a happy creature. In an early notebook, when he worked to analyze the pleasures and beauties of life, such as the "splendor of light," seeing a river's "serpentine lines narrow in the distance," and birds singing, he also listed a mysterious phrase: "virtuous happiness." If doing good is tautologically feeling good for Darwin, whether the doer is human or canine, he may have meant the happiness of acting virtuously as per biophilia.

SEXUAL SELECTION: WHO CHOOSES WHOM AND HOW?

Viewed through the lens of biophilia, sexual passion serves two goods: reproduction (genetic survival) and physical pleasure. In the view of Darwin and today's evolutionary biologists, the two are intimately linked through the passion of parental love. When he displays interspecies parental caretaking as a triumph of instinct, he hails the power of the physical nurturing and social-bonding reflexes. He argued that for the "social animal," tactile contact was emotional connection. At first, he speculated in his notebook: "How does Social animal recognize and take pleasure in other animals . . . by smell or looks. But it does not know its own smell or looks.—No doubt it may be attempted to be said that young animal learns parents smell and look & so by association receives pleasure." When he refined this thought in *The Expression of the Emotions in Man and Animals*, he wrote: "Dogs, when feeling affectionate, like rubbing against their masters and being rubbed or patted by them, for from the nursing of their puppies, contact with a beloved object has become firmly associated in their minds with the emotion of love." Touch meant love; body and heart were one for Darwin, ever attuned to the raw instincts that increase life.

Darwin's factless understanding of the origins of sexual love was proven correct: Animals raised by humans arrive at sexual maturity sexually attracted to humans. This has been documented among geese and assorted birds, as well as pandas and dolphins. Nurturance is the basis of sexual attraction. That sexuality emerges from the sanctity of parental care surely seemed abominable in Victorian culture, as it would in societies that define sexuality as destructive. Even scientists looked askance at Darwin's posited entanglement between what they saw as opposites. With such ideas, Darwin continued to pour salt into the cultural wounds he had inflicted by describing how our morality descends from animals, not a Western God.

The evolutionary import of sexual reproduction was a subject Darwin treated in part 2 of *Descent*, called "Selection in Relation to Sex." In this segment of his opus, he spelled out an innovative theory he termed "sexual selection," which he built, in part, to explain matters left mysterious by the theory of natural selection in *On the Origin of Species*. Sexual selection theory explains behaviors and physical morphologies aimed at reproduction, offspring, the perpetuation of genes, whatever the term—in contrast to simple survival of the self. For Darwin, sexual selection was crucial in understanding how two physically divergent genders evolved in one species. Eventually he declared in *Origin* that sexual selection was a stronger evolutionary force than natural selection. The facts and principles of sexual selection have since been documented across multitudes of species by twentieth-century scientists, with the controversial exception of *Homo sapiens.*

Sexual selection theory has two main components, as noted previously: competition for mates and selection of mates. Selection takes place in organic beings from hummingbirds to emus: females base their choice of mates on the "charms" of males, who "delight to display their beauty," as Darwin put it. In "Selection in Relation to Sex," he lavished detail on the physical and behavioral charms of importuning males across species who issue forth in mating season with courtship dances, elaborate nests, and virtuoso songs. Other male features that females do not sport, which he called "ornaments," are in fact weapons. But whether for war or "beauty," males flaunt in sexual display fantastical wings, protuberances, claws, jaws, building skills, gymnastics, and other tools. Darwin's natural selection theory could not account for such disparity between males and females.

He went on to observe that ornaments used by males solely during mating season "have been acquired in not a few cases at the cost of increased danger from enemies" or a decreased ability to survive, such as the peacock's ungainly fan and the buck's awkward antlers. Both are handicaps that compromise their bearers when chased by their predators; females flee better. He focused, as noted, on the feathers: "the plumes and other ornaments of the male must be of the highest importance to him;

and . . . beauty in some cases is even more important than success in battle." Darwin is saying that seductive power and reproductive success count more than surviving battles, as if the urge for life after life is stronger than the fear of death in this life, which would again reflect the very potent biophilia of the selfish gene.

Darwin argued that the female of a species determines the development of odd male morphologies because they signal health, strength, and resources for offspring. Plumage, antlers, decorated nests—these are energetically expensive for the male, requiring much effort. So one implication of his theorizing on these matters was that it gave unprecedented power to females, the power over the life, death, and very shape of males. Contemporary society recoiled over sexual selection—the modern evolutionary theorist Mark Ridley summarized the cultural and scientific reaction of Darwin's day: "any theory it seemed was preferred to the idea of female preference for male beauty."

Moving from beauty to violence, Darwin offered one of his paradoxes: "The season of love is the season of battle." He described the male competitions for females as not only requiring the use of distinctively male morphologies but also as requiring "pugnacity." It is another of his favorite words. He was fascinated to note the same pugnacity in the females of those rarer species in which females compete and fight for males. In *Descent*, he informed readers that in species in which females do the "courting," they become the stronger, more aggressive, more "pugnacious" sex. He introduced an emu in which the female is larger, while the male incubates, raises, and defends the young. Here, he added, "we have a complete reversal not only of the parental . . . instincts, but of the usual moral qualities of the two sexes: the female being savage, quarrelsome and noisy, the males gentle and good." Likewise the female painted snipe is larger and more ornate than the male and "has acquired an eminently masculine character," he wrote.

We do not need to name all the notorious species in which females cannibalize their mates so as to fortify themselves for motherhood. But the phenomenon plagued Darwin. In trying to explain the female spider who eats her baby-daddy after they mate, he wrote: "The carcase of her husband no doubt nourishes her, and without some better explanation . . . we are thus reduced to the grossest utilitarianism, compatible, it must be confessed, with the theory of natural selection." Right again. Scientists do not necessarily like their discoveries. Once, when Darwin was disappointed to learn a hypothesis-quashing fact about how flowers bud, he wrote to Huxley that he would "not allude to this subject, which I rather grieve about . . . but, alas! a scientific man ought to have no wishes, no affections—a mere heart of stone." What is true is true, he felt, whether we like it or not.

IN *HOMO SAPIENS*

When the issue was sexual selection among *Homo sapiens*, Darwin waffled. He had portrayed the animal world as largely full of preening and pugnacious males. These males exercised their voices, bodies, and skills to attract females, and they were disproportionately and self-destructively ornamented and weaponized. It seemed a contrast to *Homo sapiens*, who for centuries and across cultures has thought of and seen women as the ones preening and men the ones choosing. Darwin's view of the human dynamic was complicated and contained the kinds of contradictions he was capable of. A fellow biophile would say he took an unflinching look at contradictory and unsettling facts. Put less charitably, he vacillated like a wrecking ball.

On the one hand, he saw phenomena indicating that women choose men, the predominant method in other animal species. He felt that, accordingly, the same pugnacity that describes the males of species in which the female chooses also describes human males. He noted in *Descent* that "savage" men are the more decorated gender, doing the preening, luring the women. He stated simply that women choose among "the more attractive men." This scenario holds that men are the ones doing the luring, preening, and competing for the sexual favors and reproductive assets of women, and this in turn gives women the prerogative of choice and, over the millennia, control of the morphology of men's bodies. Darwin wrote, "I conclude that of all the causes which have led to the differences in external appearance between the races of man, and to a certain extent between man and the lower animals, sexual selection has been by far the most efficient." More generally, Darwin wrote, "The exertion of some choice on the part of the female seems almost [a] general law." Of *Homo sapiens*, he likewise wrote that "women . . . generally choose not merely the handsomer men . . . but those who were . . . best able to defend and support them. Such . . . pairs would commonly rear a larger number of offspring."

On the other hand, Darwin saw that in many human societies, unlike most other animal societies, men choose women, with women doing the luring: "Women are everywhere conscious of the value of their beauty; and when they have the means, they take more delight in decorating themselves with all sorts of ornaments than do men." Also in accord with the idea that men choose, he pointed out that in *Homo sapiens*, women are stronger and more physically resilient, noting that male infants die more frequently than females, are less hardy, and are more frequently born with deformities. Today biologists have confirmed that while men have larger and more muscular bodies than women in general, women are indeed constitutionally stronger, far more likely to withstand disease, starvation, severe weather, and other adverse conditions. This would indicate, by Darwin's theory, that women are the stronger ones and therefore doing the luring.

He also showed women adopting male-animal-style weapons of charm. He wrote that when seeing "male birds elaborately displaying their plumes and splendid colors before the females . . . it is impossible to doubt that the females admire the beauty of their male partners. As women everywhere deck themselves with these plumes, the beauty of such ornaments can not be disputed." Darwin surely would have enjoyed the fact that the tool—the feather—made a trans-species and transgender hop to serve one need.

Singing was another male-animal-style weapon of charm Darwin saw employed by female *Homo sapiens*. He offered the "inference" that women's sweeter voices and better musical powers are "acquired . . . in order to attract the other sex." He acknowledged too that men are susceptible to a pretty face. "In civilized life man is largely, but by no means exclusively influenced in the choice of his wife by appearance," he asserted in *Descent.* On this view, a man chooses a woman who is pretty as a peacock and sings like a nightingale. Darwin pointed out too that women are more beautiful than men, attributing this morphological divergence to sexual selection by men.

Altogether, it seems for Darwin, human sexual selection goes both ways: men lure and women choose, *and* women lure and men choose. He was never afraid of this kind of contradiction or bothness, a comfort with paradox that is part and parcel of biophilia (see chapter 1). He knew evolutionary theory often lacked either-ors and their comforts. His biophile's sense of truth enabled him to rise past the conventional truisms of the flawed gender norms of his day.

NATURE'S RAINBOW

Darwin's work also collapsed the heteronormative order by wrecking the idea of two complementary and discrete sexes. He knew, in the depths of his sense of life and with his legions of deconstructing facts, that the male and female binary was not a God-given structure: He "taught that man was evolved woman," as the historian of science Cynthia Russet summed up. In Judeo-Christian lore, in the first story of the Bible, Eve was built with Adam's rib. In the very first chapter of *Descent*, Darwin told it the other way around: the female is nature's foundational sex. He found evidence in the defunct uterus in the male of many a species. Darwin also pointed out, "It is well known that in the males of all mammals, including man, rudimentary mammae exist. These in several instances have become well developed and have yielded a copious supply of milk." Today, we know that human fetuses must get just the right dose of just the right hormones at just the right time to become a full biological man. Right again. And the idea of women as primary was a far cry from the thesis of a nineteenth-century *Popular Science Monthly* article that questioned the biological necessity of women.

Darwin had begun problematizing the idea of two sexes in *Origin* with that scandalous assertion: "At a very early embryonic period both sexes possessed true male and female glands. Hence some extremely remote progenitor of the whole vertebrate kingdom appears to have been hermaphrodite or androgynous." He recognized organic beings that are half-male, half-female, as well as hermaphroditic plants. He continued the demolition of the two-sex model in *Descent*'s alternative sexual pecking orders and parental-care arrangements among the birds and the bees. He catalogued males acting as mothers with a full complement of nurturing behaviors: Male syngnathous fish carry their young, he pointed out, and today's biologists know of species whose males carry young in diverse parts of their bodies at all stages of gestation and postnatal care, such as the pipefish, seahorse, and nurseryfish. He mentioned bird species in which males brood eggs and nurse young. He was as well aware of the natural vagaries of gender identity as the thriller writer and doctor Michael Crichton, whose *Jurassic Park* features as a doomsday-plot device the ability of frogs to control whether and when they are male or female. Here, Darwin's work demolished yet another binary that seemed a verity: man and woman are not opposite. In telling us that man was born woman, in his perception of the facts of life, he stirred up as much gender trouble as any latter-day feminist.

Darwin's biophilia led him to sense, analyze, and forward, once again, a new and disruptive truth, this time about the polyvalence of genders. The modern linguist and biogeographer Bruce Bagemihl has catalogued biology's alternatives in a database of 450 species that includes "animals with females that become males, animals with no males at all, animals that are both male and female simultaneously, animals where males resemble females, animals where females court other females and males court other males." In sum, he wrote, "Many animals live without two distinct genders, or with multiple genders." Darwin would no doubt relish the rainbow of variations.

By the time Darwin was done with sexual selection theory, men were not the primary sex, nor the default sex, nor even much a separate sex. Males were not always stronger and braver than females. Males are not supreme across the animal "kingdom," its name notwithstanding. He told us that males are mostly beggars and not choosers in the mating game. He had asserted that the preferences of women controlled the destiny of men's bodies and the nature of his mind. He decoupled dominant man and submissive woman. His biophilia helped him see through the frosted-glass ideology of many a culture, and he had seen many.

THE JOY OF LIVING

One more variety of passionate experience Darwin treasured and studied was joy. He was a student of joy in children and dogs. Playfulness is the easiest way to see joy, of course; it is a burbling kind of biophilia. He told his readers that happiness is "never better exhibited by young animals, such as puppies, kittens, lambs, etc, when playing together, like our children."

Darwin was not shy about his love for joy and the play of "animal spirits." His perfect example of joyful playfulness was the dog shown in this passage from *Expression*: "A dog, in the highest spirits, careers like a mad creature round and round his master in circles, or in figures of eight. He then acts as if another dog were chasing him. This curious kind of play, which must be familiar to everyone who has attended to dogs, is particularly apt to be excited, after the animal has been a little startled or frightened, as by his master suddenly jumping out on him in the dusk." Darwin was, of course, that anonymous master. But Darwin is wrong here: Only a person who knows dogs can elicit such delirious romping. Only a person who has "attended to dogs" with every cell of their heart and who has an affinity for play. Only a person like Darwin, able to launch those "highest spirits" so as to play hide-and-seek just for the joy of it. Darwin was convinced that "dogs laugh for joy."

But Darwin was not a parent who let the dogs roam free while leashing his children to their desks—he valued joy and play in his children. Francis recalled his father's way with small children, the "especial pleasure in the games he played with us." He was a "delightful play-fellow," wrote his daughter. She wrote:

"It is a proof of the terms on which we were, and also of how much he was valued as a play-fellow, that one of his sons when about four years old tried to bribe him with sixpence to come and play in working hours." His pleasure in their play, in the sight of life lifing, was as plain to his children as the nose on his face. (See chapter 7 on Darwin's marriage and domestic life.)

Perhaps ironically, the measure of Darwin's love of joy can be taken by the depths of his sorrow when his daughter Annie died at age ten. Annie was uniquely full of joy, as he told it in his own private eulogy. He described how "joyousness and animal spirits radiated from her whole countenance. . . . The spirit of joyousness rises before me as her emblem and characteristic; she seemed formed to live a life of happiness." For Darwin, her joy was her virtue, just as "joy is virtuous" in Fromm's definition of biophilia.

Moreover, for Darwin, virtue was also her joy, living proof of the pleasures of altruism. He wrote, "Her dear face now rises before me, as she used sometimes to come running downstairs with a stolen pinch of snuff for me, her whole form radiant with the pleasure of giving pleasure." Darwin wept over her clandestine gift and her joyful benevolence. "In the last short illness [she] never once complained; never became fretful; was ever considerate of others. . . . When I gave her some water, she said, 'I quite thank you;' and these, I believe, were the last precious words ever addressed by her dear lips to me."

As noted earlier, Darwin once scrawled a note to himself about "virtuous happiness," though what he meant is a mystery. But it seems to define Annie, in how her joy and her virtue wove together, the warp and woof of one cloth, one instinct, biophilia.

Chapter Six

BEAUTY IS LIFE, LIFE BEAUTY

Beauty rests on necessities.

—Ralph Waldo Emerson

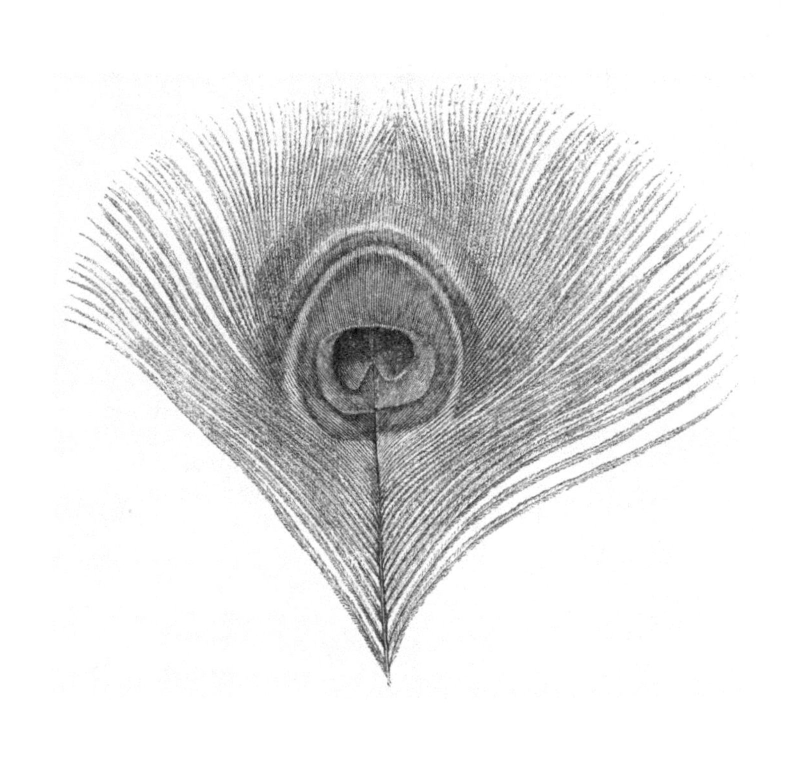

FIGURE 6.1 Birds of a feather.

Darwin was initially befuddled by the beautiful plumage on male birds such as the peacock, Argus pheasant, bird of paradise, and widow-bird of southern Africa. Their fine feathers "trouble them during a high wind . . . [and] render their flight heavy," and their "beauty was a source of danger," Darwin noted. But those very feathers aided the males' reproductive success. The "bright colors," "magnificent train[s]," and "various ornaments" decorating males were selected for by "the females having preferred . . . more highly ornamented males," he explained. Regarding the peacock, he wrote, "Many female progenitors . . . must, during a long line of descent, have . . . unconsciously, by the continued preference of the most beautiful males, rendered the peacock the most splendid of living birds." And the peahen expressed herself, Darwin continued, always making the "first advances." What did all this mean? Darwin wrote bluntly in a letter, "It is an awful stretcher to believe that a peacock's tail was thus formed; but, believing it, I believe in the same principle somewhat modified applied to man." Implying that woman thus modified man, Darwin's theory of sexual selection challenged gender norms—but that is an entirely different story. As for appreciating a feather's beauty, it is a cross-species trait, Darwin pointed out, noting that both genders in a range of human cultures adorn themselves with feathers, as if echoing an instinct about their prowess.

Source: Reproduced with permission from John van Wyhe, ed., *The Complete Work of Charles Darwin Online*, 2002–, http://darwin-online.org.uk/.

A PREDICTION from the poet Wallace Stevens: "They will get it straight one day at the Sorbonne. / We shall return at twilight from the lecture / Pleased that the irrational is rational." We often think of beauty as irrational: we do not require that sculptures or babbling brooks make sense. We may define beauty as defying logic and language. But had Darwin lectured at the Sorbonne, he might well have met Stevens's ideal, explaining how the sense of beauty ensues from the unrelenting need for survival, the primal biophilia, how it developed from our ancestors' responses to obtaining the essentials that ensure life. He might have explained how putatively irrational art is the pragmatic child of the most inexorable rigors. In his book *Art as Experience*, the twentieth-century philosopher John Dewey wrote of the "continuity between the refined and intensified forms of experience that are works of art and the everyday events, doings, and sufferings that . . . constitute experience. Mountain peaks do not float unsupported; they do not even just rest upon the earth. They are the earth in one of its manifest operations." For Darwin too, a love of beauty is an upwelling of the drive for life. His biophilia, of course, enabled him to see this particular aspect of biophilia. Darwin might have titled his Sorbonne lecture "On the Origin of Beauty by Means of Survival."

BEAUTY IS AS BEAUTY DOES

The endless forms of life were the definition of beauty for Darwin. Watch him rave in his travel diary about the Brazilian rainforest:

> Humboldt's glorious descriptions are & will for ever be unparalleled: but even he with his dark blue skies & the rare union of poetry with science which he so strongly displays . . . falls far short of the truth. The delight one experiences in such times bewilders the mind,—if the eye attempts to follow the flight of a gaudy butter-fly, it is arrested by some strange tree or fruit; if watching an insect one forgets it in the stranger flower it is crawling over,—if turning to admire the splendour of the scenery the individual character of the foreground fixes the attention. The mind is a chaos of delight, out of which a world of future & more quiet pleasure will arise. . . . [One day later:] The day has passed delightfully: delight is however a weak term for such transports of pleasure: I have been wandering by myself in a Brazilian forest: amongst the multitude it is hard to say what set of objects is most striking: the general luxuriance of the vegetation bears the victory, the elegance of the grasses, the novelty of the parasitical plants, the beauty of the flowers.—the glossy green of the foliage, all tend to this end.—A most paradoxical mixture of sound & silence pervades the shady parts of the wood,—the noise from the insects is so loud that in the evening it can be heard even in a vessel anchored several hundred yards from the shore.—Yet within the recesses of the forest when in the midst of it a universal stillness appears to reign.—To a person fond of natural history such a day as this brings with it pleasure more acute than he ever may again experience.

Note that he foresaw that this panoply would abide in his mind, which is, as noted earlier, an example of his intuitive fact-knowing, inasmuch as this region is the most biodiverse in the world, an apex, in fact. In the biophilia he shares with Edward O. Wilson, Darwin was attuned to nature by nature, one reason for his taking the arduous trip that brought him to Brazil. "When we talk of higher orders [of organic beings], we should always say, intellectually higher. But who with the face of the earth covered with the most beautiful savannahs and forests dare to say that intellectuality is only aim in this world?" he asked himself. Natural beauty was his metric when he judged human art: "With a book as with a fine day, one likes it to end with a glorious sunset," he wrote in a letter to Huxley.

Music was the human art that most moved Darwin. A college friend of his told Francis of listening to music in a church with him: "Accompanying him to the afternoon service at King's . . . we heard a very beautiful anthem. . . . He turned round to me and said with a deep sigh, 'How's your backbone?'" Francis continued, "He often spoke in later years of a coldness or shivering in his back on hearing beautiful music." Music is close to nature. As Darwin told us, animals' "vocal and instrumental sounds . . . so commonly serve as a love call or as a love-charm, that the power of producing them was probably first developed in connection with the propagation of the species." Human-made music is a direct descendant of birdsong courtship. It has an umbra of sexual pleasure and a legacy of reproductive value. Music is also temporal, as is life. And its connections with mathematics and physics, which are both natural sciences in their own way, have been explored by many. All reasons why music might be the art of choice in an evolutionist's biophilia.

Music retained its power for Darwin throughout his life, though as he aged, he wrote, its role changed. No longer spine chilling, "Music generally sets me thinking too energetically on what I have been at work on, instead of giving me pleasure." Later in life too, he lost his taste for Shakespeare. Bemoaning his waning "aesthetic tastes," he speculated that the loss "may possibly be injurious to the intellect, and more probably to the moral character, by enfeebling the emotional part of our nature." Between these lines I read an assumption that beauty strengthens us; it is practical. Perhaps human practices of beauty drew him less as he got older because he still saw so much natural beauty to explore and explain, and every day he saw less time left. His son Leonard wrote, "His appreciation of natural scenery remained quite undimmed to the end of his life." Darwin wrote in *On the Origin of Species*: "Natural Selection . . . is as immeasurably superior to man's feeble efforts, as the works of Nature are to those of Art." As beautiful as Shakespeare is, his truths are not facts. Life is beauty for Darwin, and beauty life. That is what becomes of a singular biophilia.

THE ORIGIN OF AESTHETICS

Utility is the hobgoblin of aesthetic theory. Many scholars have expended many valiant words trying to banish use from the provinces of beauty. One premise they often share is that beauty is by definition useless, paradigmatic examples being landscape paintings, diamond tiaras, and other objects that exist solely for aesthetic pleasure, consume resources without tangible benefit, etc. But for Darwin and evolutionary biologists, as previously noted (chapter 5), it is a truth universally acknowledged that the ornaments that males of so many species flash in mating displays demonstrate their health and strength. The peahen who chooses the peacock with the iridescent feathers that telegraph health (see chapter 3) displays a "sense of beauty" that reflects her common sense, or, more precisely, her instinct to choose the healthy male. Her aesthetic preference has utility.

In another of Darwin's interspecies leaps, he connected the utilitarian displays of such creatures as peacocks to "human" displays: "As with the artificial ornaments used by savage and civilized man, so with the natural ornaments of birds, the head is the chief seat of decoration." Think antlers and crowns, tusks and wigs and hair dye. And if among other animals beauty usefully serves biological ends, then it is a substrate for the human aesthetic sense, which has evolved and diversified, as organisms and phenomena do. The modern evolutionary theorist John Tyler Bonner writes, "We can find the seeds of human culture in very early biological evolution." When physical "beautiful" signals lead to reproduction and enable the healthiest, most resource-rich signalers to perpetuate their genes, we see beauty serving life. And when Fromm tells us that in the "ethics" of biophilia, "Good is all that serves life," we can see beauty's utility in biology as a facet of biophilia.

Thus all organic beings, simple and complex, are hardwired for beauty in Darwin's view: "The perceptive powers of man and the lower animals are so constituted that brilliant colors and certain forms, as well as harmonious and rhythmical sounds, give pleasure," he wrote. He felt that all animals have an "appreciation of the beautiful in sound, color, or form." He amassed a cross-species depth and breadth of aesthetic preferences. Shining feathers, elaborate bowers, huge horns, long songs, cozy caves, expansive views—all these guide courtship behaviors, mating choices, and reproductive success. Darwin plumbed the theoretical mechanics of animals' aesthetic bells and whistles in his theory of sexual selection. He told his readers flat out in *The Descent of Man, and Selection in Relation to Sex* that beauty has evolutionary utility. He earned scorn in his century for arguing that animals possess a sense of beauty and that it figures in their reproduction. But again there is that refrain: Research has proven him right.

He allowed that his assertion that animals possess aesthetic sense was "astonishing in the case of reptiles, fish, and insects. But we really know very little about the minds of the lower animals. It cannot be supposed that male Birds of Paradise or Peacocks, for instance, should take so much pains in erecting, spreading, and vibrating their beautiful plumes before the females for no purpose." On this view, the philosopher Edmund Burke commits an anthropocentric error of reasoning when he dismisses the utility of beauty on the grounds of what he sees as the ugliness of the "wedge-like snout of a swine, with its tough cartilage . . . the great bag hanging to the bill of a pelican," and the hedgehog's "prickly hide." The bigger the great bag of the pelican, the more food it can scoop up, and the better fed are the pelican's little ones. And the pricklier the hide of the hedgehog, the more likely that it will survive an attack and live another day. The attractiveness of such attributes is in the eye of conspecific female beholders. It is for neither Burke nor us to judge.

Darwin speculated that for *Homo sapiens*, "beautiful" objects are things that enable us to live and reproduce, as when he told us our love of music descended from the mating call. On the subject of sex and beauty, he was explicit when musing to himself in a notebook, "Some forms seem instinctively beautiful, as round ones. . . . Stallions licking udders of mare strictly analogous to men's affect[ion] for women's breasts." He also wondered, "Western men" have "one notion of beauty and negroes another; but it does not explain the feeling in man." Where the sense of beauty came from was a question that plagued Darwin, but he knew that "many . . . cases could be advanced of organs and instincts originally adapted for one purpose, having been utilised for some quite distinct purpose." And so a "notion of beauty" may arise from "instinct." Beauty has its own character distinct from survival, of course; again, body becomes mind. After Darwin, the philosopher George Santayana, the psychoanalyst Sigmund Freud, and legions more placed sex at the root of aesthetic feeling, figuring sex as the raw and aesthetics the cooked. As Darwin had scribbled in a notebook, "pleasure in the beautiful distinct from sexual beauty is acquired taste"—that is, acquired through biophilia.

Some beauty that we see descends from the pleasure of obtaining our basic needs for life. Certain natural phenomena critical to survival are considered beautiful by people of many cultures. The trill of a brook is universally considered beautiful, and it signifies the pure water necessary to life. A fire gives sustaining warmth. Rain sluicing down the roof is the sound of shelter. An expansive view affords safety from enemies. The cool of a summer evening is relief from heat: "The freshness of night has been fresh a long time," Wallace Stevens pointed out. When leaves susurrate in a breeze, that is fresh air made audible.

These phenomena enable us to live, so we feel them as beautiful. As Darwin's grandfather asserted in his book *Zoonomia*: "Our perception of beauty consists in our recognition . . . of those objects . . . which have inspired our love by the pleasure, which they afford to many of our senses, as to our sense of warmth, of touch, or smell, of taste, hunger and thirst." George Santayana echoed the thought: "Beauty is pleasure objectified." Beauty is almost tautologically pleasure. What is culturally universal is biologically rooted; these preferences are called "living fossils" in at least one modern theory. Research in evolutionary psychology today has revealed that a tree of a certain shape—the wide umbrella of the hardy African acacia, which offers maximum shade in the savannah where our ancestors lived—is universally considered the most beautifully shaped tree. The researchers, John Tooby and Leda Cosmides, consider that aesthetic responses are made of "cognition and affect wedded" and are so instinctive they are "not open to introspection."

In nature are the first beauties, then, the oldest aesthetic cliché. One philosopher in this tradition, Walter J. Bate, defined art as "an imitation of what is essential in nature." The modern philosopher Suzanne Langer extended the theme, adding a twist of evolutionary psychology: "The emotive content [in art] is apt to be something much deeper than any intellectual experience, more essential, pre-rational, and vital, something of the life-rhythms we share with all growing, hungering, moving, and fearing creatures." Biophilia is one of those life-rhythms.

All kinds of theories—evolutionary and philosophical and psychological—trace a path from animal survival to aesthetic sense. As the modern evolutionary theorist Geoffrey Miller has pointed out, Darwinism is a "universal acid . . . dissolving the cultural into the biological," a juggernaut project that Darwin initiated willy-nilly when he whispered in the privacy of a notebook, quiet as the flap of a butterfly's wings, "mind is function of body." We have since then had the nineteenth-century psychologist William James, with his theories of the embodied mind, and Freud, again, with his stories of instincts that find new venues to satisfy ancient impulses. These are but a few of the theories positing a flow from body to mind that accord with an origin of the love of beauty in the drive for life.

All these theories that bring a desire for life from body to mind trace the same pattern that ran through Darwin's thought when he jotted, as already noted, in a notebook, "Instinct is a modification of bodily structure . . . and intellect is a modification of instinct." He constantly conflated physical and mental states: "Think, whether there is any analogy between grief & pain—certain ideas hurting brain, like a wound hurts body—tears flow from both. Joy a mental pleasure. . . . The shudder of pleasure of music." He saw continuity between the visceral and the ineffable, that physical satisfaction is the model for the mental pleasure of beauty. This passage from body to mind also recurs in philosophy. For example, Alfred North Whitehead expounded a philosophy of organism in which "conceptual feelings pass into the category of physical feelings . . . conversely, physical feelings give rise to conceptual feelings." The philosopher Edmund Burke asserted that "the senses are the great originals of all our ideas."

Survival as the origin of the aesthetic sense may be corroborated by the fact that all aesthetic experiences elicit just one response. Studies of human aesthetic pleasure have found that people experience a distinctive brain state when sensing their own divergent ideas of beauty. So whether viewing sculpture, reading poetry, hearing music, scenting flowers, learning a mathematical theorem, stroking a cat, even receiving a friendly smile—all evoke one singular sensation in the brain, in a tiny sweet spot called the medial orbitofrontal cortex. It is as if we have only one urge to satisfy. This hub in the brain's reward and pleasure center was cultivated by the primal pleasures of ongoing life. It is more ancient than the arboreal quadruped. But scientists were not the first to find it, as the poet Samuel Taylor Coleridge told us: "The safest definition, then, of Beauty, as well as the oldest, is that of Pythagoras: The Reduction of Many to One."

The binary beauty/survival must be added to the list of others Darwin demolished—nature/culture, male/female, animal/human, morality/instinct, and cause/effect. Evolutionary theory takes all of us—and all of our ideas—for a ride on the Möbius rollercoaster.

AESTHETIC THEORY GETS VISCERAL

If the drive for life gave birth to the sense of beauty, so too does beauty aid life, each biophilia serving the other, like M. C. Escher's two hands drawing each other. Beauty aids survival by teaching us how to field the new. *Homo sapiens* has an appetite for the new, as Darwin wrote in *Origin*: "It is in human nature to value any novelty." It is a trait of biophilia as well, as Fromm pointed out. We also need to know how to field the new because change is the only constant.

When the Darwin scholar and theorist Morse Peckham analyzed artistic behavior, he began by asking: "What is its function in biological adaptation?" He hypothesized that we create and engage with beauty and art to "rehearse" skills for comprehending the new. He speculated that we have what he called a "rage for chaos"—a phrase from Wallace Stevens—which functions "to satisfy a physiological need for a more stimulating environment than the order-directed social environment offers." Other aesthetic theorists connect art and survival in other ways that endow aesthetic activity with adaptive value. The aesthetic theorist Ellen Dissanayake has argued, for example, that the arts aid and celebrate survival skills: for example, cave paintings that chronicle hunting. In her novel study *Homo Aestheticus*, she wrote that "the arts are extensions of what we have evolved to do naturally in order to survive and prosper"—exactly the business of biophilia. And the evolutionary psychologist Geoffrey Miller recently argued that engaging in the arts is linked with reproductive success.

In this vein, we can also find the modern evolutionary psychologists John Tooby and Leda Cosmides elaborating on the import of Picasso's famous quip that "art is a lie which makes us see the truth." They posit that aesthetic experiences serve adaptation by "unleashing our reactions to potential lives and realities." They point out: "To call art 'lies' simply acknowledges that a simulacrum of individual experience has been manufactured largely out of false propositions or orchestrated appearances. . . . The truth inheres in what the experience builds in us." And knowing the truth is, aside from being a desideratum in biophilia, is certainly of survival value.

So life creates beauty, and beauty creates life: yet another Möbius. The modern theory of gene-culture coevolution holds, as noted earlier, that "culture is created and shaped by biological processes while the biological processes are simultaneously altered in response to cultural change," as Charles Lumsden and Edward O. Wilson write. We do not see or feel ourselves evolving through culture, of course, but neither do we see or feel Earth's spin. Nevertheless, to return to the philosopher Peirce, the physics of the universe shapes our minds and our lives (see chapter 1). Peirce premised, as noted earlier: "Our minds having been formed under the influence of phenomena governed by the laws of mechanics, certain concepts entering into those laws become implanted in our minds." Such as paradox. So even if the Möbius is a paradox, it is a natural form, and it is natural that we evolve along its curves.

Furthermore, the resemblance between the sense of beauty and the process of evolution is striking. It is rooted, as ever, in the physics of the universe. The cosmologist John Barrow has explained in *The Artful Universe* that two laws govern the "Universe's deep structure," encompassing the "constants of Nature [which] enforce uniformity and simplicity, while initial conditions and symmetry-breakings permit complexity and diversity." More simply, there are two forces in the universe: those creating uniformity and those creating variety. This pair figures precisely in Darwin's definition of evolution. He explains in

Origin : "These elaborately constructed forms . . . have all been produced by laws acting around us. These laws . . . being . . . Reproduction; Inheritance which is almost implied by reproduction; [and] Variability . . . and as a consequence, Natural Selection."

In the natural world, how do uniformity and variety interplay? Survival happens through uniformity in many matters—genes, weather conditions, food sources, and social signals. Survival equally happens in concert with variety—competing species, mutations, or geological or meteorological change. A short-beaked bird species perfectly adapted to its ecological niche may die during a sudden drought when it can not drink from a shrunken waterhole, while the rare chick born with an abnormally long beak might survive this contingent event and spark a new species that will inherit the waterhole. The species exists or evolves based on the balance of uniformity and variety. This duality is broad and deep, as the philosopher Daniel Dennett writes: "For evolution by natural selection to occur . . . you must have just the right sort of order, with just the right mix of freedom and constraint, growth and decay, rigidity and fluidity."

And given Peirce's idea that the laws of nature become implanted in our minds, perhaps it is not surprising that beauty is often defined as a mix of uniformity and variety, the old and the new, the familiar and the strange. In surveying aesthetic theories from the Renaissance though the twentieth century, the psychobiologist David Berlyne listed thinkers for whom "the conditions making beauty or aesthetic pleasure have focused on two mutually counterbalancing factors." One such thinker was the eighteenth-century philosopher Francis Hutcheson, who wrote: "The Figures which excite in us the Ideas of Beauty seem to be those in which there is Uniformity amidst Variety." The seventeenth-century philosophical luminary René Descartes stated that the object "most agreeable to the soul" must neither be too "easy to become acquainted with" nor so "difficult that it makes the senses suffer in striving to become acquainted with it." These ideas echoed in our era when Morse Peckham analyzed art as a beneficial cognitive link between order and chaos. And again when the psychologist Sylvan Tomkins wrote: "A symphony is written and appreciated as a set of variations on a theme. The enjoyment and excitement of such experience depends upon the awareness of both the sameness and the difference."

An ideal balance between uniformity and variety, a certain ratio, is necessary to the process of evolution and the sense of beauty, as in the theorist John Dewey's "stable, though moving equilibrium," which must be preserved as change happens. Modern evolutionary theorists have a term for the maintenance of mutual adjustment, the "attuned ratio," and describe how it remains precise and in equilibrium, notwithstanding change. It is also dubbed the "Red Queen phenomenon," after the character invented by the logician Charles Dodgson, aka Lewis Carroll, who told Alice that "it takes all the running you can do, to keep in the same place." Red Queen theory describes how a species must change with the times to preserve its place. The relative adjustment is in constant recalibration through natural selection. It is interesting that a logician perceived a paradox that later was discovered in the evolutionary process.

This same need for calibration figures in the sense of beauty, as when Darwin's grandfather Erasmus wrote that "beauty has been defined by some writers to consist in a due combination of uniformity and variety" in a "precise proportion." The due combination is a long-standing feature in definitions of beauty, as when Gottfried Wilhelm Leibniz asserted that in feeling beauty, we "obtain as much variety as possible but with the greatest order that one can."

Peckham pointed to a progression, a constant calibration, in the "due combination" in the experience of beauty. Peckham called on the examples of Picasso and Braque to illustrate how what people initially experience as strange becomes beautiful. "It is an old story that Bach, Beethoven, and Wagner were 'hard to hear' in their own time," remarked the philosopher Suzanne Langer. The new and unusual, seen or heard again and again, loses its novelty, passes a mental threshold into the almost usual, into the due combination of the familiar and the strange—and is experienced as beautiful. This evolves further. Peckham argued that with more exposure to a work of art, our aesthetic threshold may shift. He wrote, "A work of art that moves me today may bore me tomorrow and often does." The new gets old. What is mesmerizing may become boring. What is beautiful may become a cliché. What gives aesthetic pleasure can evolve just like an organism. Dewey summed up the progression: "Changes interlock and sustain one another."

Biophilia needs, seeks, and creates the beautiful and the new.

LOGICAL, NOT RATIONAL

A small legion of theories traces the influence of evolution on art and of art on evolution, with names such as bioaesthetics, biocultural theory, biopoetics, biosemiology, evolutionary psychology, neuroaesthetics, psychoaesthetics, psychobiology, selectionism, semiobiology, sociobiology, and species-centrism. All this work elucidates the "close relationship between aesthetics and selective advantage, between beauty and viability, between the arts and life," as one practitioner in this area, Brett Cooke, recently wrote.

Is it rational to think the visceral is the spark of all we hold most cultural? *Homo sapiens* is the much ballyhooed rational animal. But reason is a limited instrument, as Darwin kept repeating. As noted earlier, George Santayana chimed in on this point: "The ideal of rationality is itself as arbitrary, as much dependent on the needs of a finite organization, as any other ideal." Evolution per Darwin may be full of paradox and not rational, but it has its own inexorable logic. In *The Artful Universe*, Barrow premises that the human mind is created "by an evolutionary process that rewards faithful representations of reality with survival." Beauty is the principle of evolution made visible.

Chapter Seven

THE ONE GREAT LOVE OF TWO TRUE MINDS

Love . . . is an ever-fixed mark.

—William Shakespeare

FIGURE 7.1 An apple not far from the tree.
Francis Darwin's childhood drawing displays botanical accuracy, artistic skill, and a perspective on the human-as-background much like that of his father. Francis was always the keen observer, also like his father. His parents valued the "sharp-eyedness of children," he recalled, remembering "vividly the intense pleasure" he felt when he impressed his father by identifying plants in winter. Francis described a childhood attuned to the sounds of nature. He made musical instruments from plants. One such "rustic pipe" was fashioned from a dandelion stalk: "I like to remember that my primaeval oboe gave me the idea" for some botanical work, he wrote. Nature was so full of the music of life for Francis that when he wrote of plant movements, he declared that in "the land of magnification, where the little looks big and the slow looks quick, we see such evidence of movement that we wonder not to hear as well as see the stream of life that flows before our eyes." And again like his father, Francis saw a continuity between plants and people. After proving that plants sense gravity, he added an anthropomorphic touch: "Gravitation is . . . a signal which the plant interprets in the way best suited to its success in the struggle for life, just as what we see or hear gives us signals of changes in the exterior world by which we regulate our conduct." Seeing such unity across nature's species demonstrates his biophilia.

Source: Reproduced by kind permission of the Syndics of Cambridge University Library. Manuscript, DAR 185: 109f22v.

EMMA WEDGWOOD rejected the hand of perhaps four or five suitors before accepting Charles's, but this particular heiress was no flibbertigibbet. Look whom she chose after all. Nor was he *the* Charles Darwin when she married him, rather more like the boy next door: they were cousins in an intermarrying clan, as was common in their sociocultural moment. To Emma's eye, Darwin's hand was haloed by the exotic appeal of the world traveler, what with his five-year jaunt around the world. He was reminiscent of the famous explorer Mungo Park, an idol of her childhood. Darwin also endeared himself to Emma in being "humane to animals," she wrote to her aunt. She liked that he was down to earth and straightforward: "He is the most open, transparent man I ever saw." Emma foresaw that his "power of expressing affection . . . will make his children very fond of him."

For his part, Darwin preferred a "natural" woman, he revealed in a letter to Emma, who was treasured in her family for her animal spirits, as Darwin was in his. Emma's fond aunt prophesied that she would "lark it through life." In some ways, Emma epitomized the "angel in the house" then idealized in England's upper class, an empty-headed creature of grace. Emma was, for example, so musically gifted that Frederic Chopin taught her. As a newlywed with a new piano in 1839, Emma wrote, "I have given Charles a large dose of music every evening." Throughout their fifty-three-year marriage, she played for him in the evenings. Her deft fingers could style your hair in the latest curls and charm your children by cutting bears, lions, and pigs out of paper.

But Emma also read through several private libraries; could translate scientific French, German, and Italian; and spoke idiomatic French. Her father believed in education for women and put good money toward the cause with his daughters. Emma wrote her own primer for teaching reading and writing to village children and her own. She neglected her clothing and let her pearl necklace get jumbled in her jewelry box. The home she and Darwin eventually made was ramshackle with nonmaterialism. One biographer went so far as to call Emma a "disorganized . . . slob." Emma was diligent only if inspired. She compounded and administered potions to cure many an illness in the countryside around her. One of Emma's prized possessions was a notebook kept by Darwin's father, containing the remedies he developed as a country doctor. Indeed, on learning of the engagement of Charles and Emma, Darwin's father told his son, "You have drawn a prize."

Emma Wedgwood was a woman of her day in placing a man at the heart of her world, but she was distinctive too. Emma had what a friend called the "gaiety . . . of her own mind." Her daughter Henrietta said Emma had a "bright aliveness," though as time wore on, she said, her mother became "serene but somewhat grave." A great-aunt described her as "the happiest being that ever was looked on," though Henrietta noted that "jokes and merriment would all come from my father." After Darwin died, Emma wrote in a letter, "I feel a sort of wonder that I can in a measure enjoy the beauty of spring." In her last year of life, Henrietta wrote, Emma "kept an extraordinary youthfulness of mind . . . almost her most remarkable quality" and was "full of energy and enjoyment. . . . Her power of living in the lives of those she cared for made her really enjoy their pleasures at secondhand, and kept many avenues to life open that are often closed to the old." She had an innate joy, an abiding love of life.

Emma did not quite "lark it through life," as her aunt predicted, with three of her ten children dying and an invalid husband at the nexus of controversy for decades. But she was animated by the same transcendent love of living that ran deep in Charles Darwin's soul. She had the psychological biophilia described by Fromm: a "person who fully loves life [and] is attracted by the process of life and growth in all its spheres." Emma's biophilia was in accord with the assertion of the American psychologist James H. Leuba, who, as mentioned earlier, wrote that "life, more life . . . is . . . the end of religion. The love of life, at any and every level of development, is the religious impulse." Emma and Charles shared that religious impulse, even as Emma was troubled by his nonbelief in her more traditional religious ideals.

The courtship of Emma and Charles was short. Emma's favorite aunt harangued her to buy some nice clothes, warning her of the social necessity to up her sartorial game. Emma and Charles skipped the honeymoon, too. But the two were graced in a marriage of true minds, united by the spirit of biophilia. They lived a life replete with truth, beauty, nature, discovery, joy, and kindness—all facets of the love of life. They had compassion for each other, camaraderie, and mutual solicitude, and they indulged in ruthless backgammon games throughout their marriage. They let passels of children run riot for decades, playing games with "howls and screams [and] one fearfully noisy game which invaded the whole house," their son George recalled. The children were rarely chastised to be quiet while their increasingly famous and increasingly ill father worked. Charles and Emma taught their children and countless cousins and young houseguests to be kind and happy, to think carefully, and to care for life. Biophilia suffused the Darwin enclave.

Their granddaughter, the woodcutter Gwen Raverat, who grew up to hang out with the bohemian Bloomsbury group, recounted in a memoir, still in print today, that she fell in love with nature when she visited the Darwin family home:

> All the flowers that grew at Down were beautiful; and different from all other flowers. Everything there was different. And better. . . . It was adoration that I felt for the foxgloves at Down. . . . This kind of feeling hits you in the stomach, and in the ends of your fingers, and it is probably the most important thing in life. Long after I have forgotten all my human loves, I shall still remember the smell of a gooseberry leaf, or the feel of the wet grass on my bare feet; or the pebbles in the path. In the long run it is this feeling that makes life worth living, this which is the driving force behind the artist's need to create.

Biophilia, plain and simple.

SAY IT WITH FLOWERS

"The earth laughs in flowers," Ralph Waldo Emerson told us. Flowers embody a range of biophilias, from their elaborate adaptations for survival to the sense of beauty they touch within us. Emma and Charles shared a love of flowers, which pervaded their eighteen-acre roost. Charles and Emma had chosen a brick house in southeastern England, in a former parsonage. The couple took possession in 1842, and the homestead became a research station, with experiments in every corner, earthworms on the piano for a time, and a state-of-the-art hothouse home to exotic orchids and insect-eating plants. Their land nurtured apple trees, beech, cherry, chestnut, fir, mulberry, quince, oak, pear, plum, walnut, and others. Down House was a haven for growth of every kind—plants, children, knowledge, joy. And flowers.

The vectors by which flowers arrived at Down House were many. Soon after Charles and Emma moved in, Darwin arranged for the same flora to be planted that had flourished under Emma's care at the Wedgwood estate, showing the very "power of expressing affection" that Emma valued in him. Another long-term supplier was Darwin's friend, Sir Joseph Hooker, head of the Kew Gardens, who sent many a specimen adding luxe and variety. A partial list of Down's floral cornucopia: coromilla, gazanias, gentian, harebells, hawthorn, ladies' finger, larkspur, lilies, little yellow rock-rose, milkwort, phloxes, portulacas, scabious, sloes, traveler's joy, verbenas, and wild orchids. Darwin had a "sacred feeling" for those orchids, Henrietta recalled, and "taught us all to have this peculiar feeling." He wrote an opus on the "contrivances" of orchids. This list of flora tended by Charles and Emma omits the kitchen gardens, tended by a lifelong servant whose produce won country-fair accolades.

A neighbor of Darwin's recounted that one of Darwin's gardeners once said: "It's a pity though that he can't find a proper occupation for himself. Just imagine: he stands and peers at a flower for several minutes. What man with something serious to do would behave in that way?" The botanical expertise in the household was such that at the age of three, the first Darwin child, William, instantly spotted the fake blossoms Darwin had plucked from one of Emma's hats, planted in situ, and pimped out in real leaves and drops of honey to investigate the bee's fraud detection. Darwin excelled at impromptu home-schooling. Another Darwin child, Francis, painted flowers as accurately as any professional illustrator and grew into a renowned botanist. Emma had learned to tend plants as a child, tutored by her uncle John Wedgwood, the founder of England's Royal Horticultural Society, and at age twenty-eight, she declared, she "took to gardening at a great rate." Decades later, she wrote to her daughter that when she felt sad, she gardened and noted "how cheering a little exertion of that sort is." Emma even tended the wildflowers that bordered a winding path Darwin walked along for daily exercise.

Just as Darwin had taken note of Emma's love of flowers, she was attuned to his: his biographer Janet Browne wrote, "Watching him bending solicitously over his flowers, she said, Emma got the feeling that Darwin would have liked to be a bee above all other species." Makes sense to want to be a bee, drinking from flowers hour after hour, day after day. We often picture bees in bright fields, as if, like sundials, they "count only sunny hours." Emma's intuition reflected a grain of truth, as Darwin admired the instincts of bees. We know now that bees are a keystone species on the planet. Darwin's biophilia resonated to that, I believe, as it had to the Brazilian rainforest's biodiversity. All this suggests that perhaps Emma served honeycombs at tea time to feed her husband's putative wish.

Along with the couple's delight in growing flowers, they took the deepest pleasure and joy in growing children. In the Darwin household, all children bloomed with joy, not just the woodcutter Gwen Raverat. Emma and Charles created a halcyon freedom for the Darwin offspring and cousins and all their friends and the children of friends. Down House was perpetual chaos. One Wedgwood relative quipped, "The only place where you might be sure of not meeting a child was the nursery." Indeed Emma and Charles's lax parental style provoked Darwin's father to "thunderstorm," Darwin wrote to Emma. Reminiscing late in life, Emma told her neighbor Louisa Nash, "When we were young, Charles and I talked over together what we should do. The house was newly and expensively furnished. Shall we make the furniture a bugbear to the children or shall we let them use it in their play?" The upshot? Nash wrote, "They agreed that they would not worry about things getting shabby."

Accordingly, Darwin's children repurposed the staircase as a gymnastics studio, used couches and chairs as trampolines and trains, and felt all the magic of science when they assisted their father. They freely invaded Darwin's study as he worked, grabbing paper, scissors, a ruler, or some other tool for play. In exchange, they smuggled in figs and snuff, treats forbidden to him by Emma and his doctors. They ensured his little joys, and Darwin tracked theirs, Henrietta recalled: "He cared for all our pursuits and interests, and lived our lives with us in a way very few fathers do." Their joie de vivre was biophilia incarnate (see chapter 6).

Darwin encouraged his children and sensed their feelings and their biophilia. His daughter Henrietta wrote: "One trifling instance makes me feel how he cared for what we cared for. He had no special taste for cats, though he admired the pretty ways of a kitten. But yet he knew and remembered the individualities of my many cats, and would talk about the habits and characters of the more remarkable ones years after they had died." Darwin was no doubt multitasking in this matter of his daughter's felinophilia, gathering cat facts. Still, his encyclopedic knowledge of her cat soap operas showed how he shared her delight in the lives she loved.

In 2013, a British newspaper announced the existence of drawings by Francis as a child, drawn on the scrap side of a draft of *On the Origin of Species*. In one, birds and a butterfly approach botanically accurate flowers, while the second, called "The Battle of the Fruit and Vegetable Soldiers," is fantastical. Both are full of detail, proportion, and whimsy. They are pictures of Francis's love of life. And they give a picture of Charles and Emma as the ideal parents envisioned by the modern psychoanalyst Phyllis Greenacre, who wrote that a "necessary condition for the flowering of great talent is the development in the young child of [a] love affair with the world." This *love affair with the world* seems yet another description of biophilia.

As in many homes in that era, childhood diseases were rife and dire in the Darwin household, which grew, shrank, and convulsed with Emma giving birth to ten children, seven of whom survived. Darwin tended all ill children as intimately as any mammalian mother and was often called on to dispense the tenderness Emma noted in him so early on. One daughter praised his bedside manner, calling him "the most patient and delightful of nurses." Sick children were treated to a nest by the fire in their father's study, she wrote: "I remember the haven of peace and comfort it seemed to me when I was unwell, to be tucked up on the study sofa, idly considering the old geological map hung on the wall. This must have been in his working hours, for I always picture him sitting in the horse-hair arm-chair by the corner of the fire." Darwin joked, wrote Francis, that he never managed to "observe accurately the expression of a crying child" because "his sympathy with the grief spoiled his observation."

But light moments were far more plentiful in the Darwin home. One maternal contribution of Emma's to the household culture was a "galloping tune" she composed. Henrietta described a frequent scene of her childhood: the piano thundering under her mothers' fingers, "the furniture pushed on one side, and a troop of little children galloping." Emma gamboled with them now and then. She rode down their indoor sliding pond, a wooden affair set on the staircase with a pitch adjustable for toddlers and daredevils. She was also what we would call today "permissive": Emma was "courageous, even rash, in what she let her children do," Henrietta recalled, for example, allowing Henrietta to "wander . . . about the lonely woods and lanes in a way that was not very safe." As for the child who wished to ride a horse bareback? Yes. As Fromm wrote, the biophile "loves the adventure of living." Emma facilitated adventure.

As parents, Emma and Charles valued high spirits and independence; Darwin complained that his son William's attendance at a school did not open his mind but closed it. During Darwin's daily constitutional, he was surrounded by childhood joy, as Francis recalled: "The Sand-Walk was our playground. . . . He liked to see what we were doing, and was ever ready to sympathise with any fun that was going on." Perhaps their joy was an inspiration to him, which would make it virtuous as well (see chapter 5) .

Informal education was a constant at Down, inspired by natural curiosity, aka biophilia. Emma's father, Josiah, had pioneered in allowing his children to learn through curiosity and inquiry; as he directed his children's governess, her pupils were to be "allowed to act in an unrestrained manner without rules and precepts. . . . The children may be taught to exercise their faculties by inducing them to answer their own questions." Darwin's contributions to his children's education were giving them hands-on research experience, modeling the mind as a dousing rod for facts and truths, and instilling a sense of wonder. And Darwin gave of himself as an impromptu teacher. Henrietta recalled, "He always put his whole mind into answering any of our questions." And he gave of his biophilia: Darwin's neighbor Louisa Nash observed, "All the children of the family were brought up with a reverent love for living things."

As to the kind of character produced by this biophilia-suffused upbringing, Darwin's granddaughter, Gwen Raverat, offered an opinion about the five Darwin sons:

> The Five Uncles [William, George, Frank, Leonard, and Horace] . . . all had . . . a warm, flexible, very moving voice; the same beautiful hands, and, of course, the same permanently chilly feet. . . . They really were the most unselfconscious people that ever lived, those five uncles. . . . In their scientific work they showed many of the characteristics of the creative artist: the sense of style, of proportion; the passionate love of their subject; and, above all, the complete integrity and the willingness to take infinite trouble to perfect any piece of work. I always used to feel that they needed protecting and cherishing, for they never seemed to me to have quite grown up. . . . I always felt older than they were. Not nearly so good, or so brave, or so kind, or so wise. Just older.

CONTINUA

Central to a love of life is understanding that life and death are two sides of a coin. The biophile's love of life is informed by the fact of death, an understanding that life and death are not opposite but coexist in a Möbius strip. As a biologist and theorist, Darwin knew that life and death are entwined. The Möbius strip likewise informed Emma's understanding of life and death.

For all her liveliness, Emma had at a young age made peace with death. As a teenager, Emma and her sister "doctored their neighbors," according to her biographer Edna Healey, and "were accustomed to dealing with births, deaths and sudden emergencies." In the face of mortal flesh, Emma learned both to hold on to and let go of life. Emma's sister wrote to Darwin within days of the death of their ten-year-old daughter Annie: "How gently and sweetly Emma takes this bitter affliction. She cries . . . but without violence, comes to our meals with the children and is as sweetly ready as ever to attend to all their little requirements."

Emma was deeply distraught when Darwin died, of course, noting only his time of death in her diary. Words were pointless. Some other events of that difficult week Emma initially logged on the wrong days, such as "summer house blown down." This loss she entered twice and crossed out twice. Grief had confused her. She never did record when the summer house was blown down. Yet on the day of Darwin's death, her daughter noted, "She came down to the drawing room to tea, and let herself be amused at some little thing, and smiled, almost laughed, for a moment as she would on any other day." This behavior was not the proverbial English stiff upper lip. Rather, for Emma, mourning and normality were contrapuntal. Darwin's death and the loss of the summer house.

Emma did not cope with grief though repression but through expression. She did not sit still and mourn but brought her grief to life. Famously, Emma collected in a writing case small artifacts from Annie's life, including relics of her character: writing implements testified to her love of letter writing; her sewing implements to her precise hand; and a few "trinkets and treasures," as a relative wrote, testifying to her magpie love of beauty. When Henrietta discovered the box decades later, she wrote, the objects revived memories with a "strange vividness." Emma's trove of the humdrum made Annie vivid not only to her sister but also for a Darwin descendent, Randal Keynes, who was inspired to write *Annie's Box: Charles Darwin, His Daughter, and Human Evolution*, a book that became the basis of a biopic, *Creation*.

In her biophilia, Emma could transmute grief into happiness, virtue, gratitude, serenity. On the death of her third child, she wrote to her sister-in-law, "Our sorrow is nothing to what it would have been if he had lived longer and suffered more. . . . With our two other dear little things you need not fear that our sorrow will last long." When her sister Fanny died, she told her favorite aunt, Jessie Sismondi, "I do not like that you should be thinking of us as more unhappy than we are. I think we all feel . . . susceptible of happiness. . . . The remembrance of her is so sweet." Then too, Emma believed in a heaven for the deserving, so she resolved to keep "my Fanny's sweet image . . . in my mind as a motive for holiness. . . . Such a separation as this seems to make the next world feel such a reality." Emma always found the silver lining, the grace. After Darwin's death, she wrote to her daughter: "Life is not flat to me. . . . I do feel it an advantage not to be grudging the years as they pass and lamenting my age."

Emma saw mercy in death. Whether accepting the need to euthanize a hardscrabble litter of kittens or taking comfort in the thought that her sister Elizabeth's death released her from her suffering, Emma "was often consoled by the thought that God, like a good gardener, knew that there was a time for the knife," as Healey wrote. Emily Dickinson once commented in a letter, "Why the Thief ingredient accompanies all Sweetness Darwin does not tell us." With the equanimity of biophilia, his wife might have affirmed that yes, sweetness *will* be stolen but will also be found. Death follows life, but life follows death; perhaps Emma's belief in an afterlife assisted, but she felt nature's infinite Möbius.

NOTHING BUT THE TRUTH

As Fromm wrote, a person with biophilia is "capable of wondering . . . prefers to see something new." If biophilia encompasses a passion to find the new in the old, the truth under the mystery, it is also a passion that keeps the mind young, as Emma's remained, in the view of her daughter Henrietta. Charles and Emma shared this passion throughout their lives. Emma resonated to the primacy of truth within her husband: "Every word expresses his real thoughts," she told her aunt. She respected his quest for new truths: Despite her sorrow and fears about the effect of Darwin's atheism on his afterlife, she wrote to him, "I . . . feel that while you are acting conscientiously and sincerely wishing & trying to learn the truth, you cannot be wrong."

Scientists love truth by definition, of course. But Darwin had an especial affinity. He collected criticisms of his theories, mulling them over, culling further truths, then revising his ideas and books; he revised *Origin* six times and *Descent* and *Expression* twice. However, there was no reply he could make to the scoffings of one contemporaneous critic, his former professor, the geologist Adam Sedgwick: "You have . . . started up a machinery as wild, I think, as Bishop Wilkin's locomotive that was to sail with us to the moon."

Emma had a knack for truth herself. Her childhood home had been filled with leading figures in politics and culture, exposing her to every kind of perspicacity. Her education helped, as did the visionary Wedgwood heritage. Her daughter said that all her life Emma had a "many-sided interest in the world, in books, and in politics." Emma applied her acumen in her marriage. For example, Charles considered Emma an excellent editor. They wordsmithed his sentences down to the commas. When Emma read Darwin's earliest draft of evolutionary theory, she flagged the term "natural selection" and scribbled "great assumption" in the margin, with the uncanny prescience of the best editors. She understood the revolution that those two plain words would wreak. Emma knew what was what.

Emma knew not only the truth of ideas and facts but of feelings. Today we would say she had much emotional intelligence. Emma's family considered that she had been endowed with her mother's empathy. One biographer described her as "sensitive to the feelings of others." Charles marveled in a notebook, "E. says she can perceive sigh, commenc[ing] as soon as painful thought crosses mind, before it can have affected respiration." Emma liked to know a person's experience, writing to her aunt, "I must say you write the pleasantest letters in the world, because you tell your own feelings and that is what one is most interested about." Emma connected to feeling; as a friend wrote: "More than any woman I ever knew, she comforted."

Sensing true feeling and a feeling for truth were common in the Darwin home. Henrietta recounted the visit of a relative who was shocked at the table to hear that Henrietta's brother

William wanted to see the carcass of a dog hit by a train; Henrietta wrote that the relative had "no understanding of boy nature, or indeed human nature." William's desire to see and know was not only acceptable conversation but a respected need for information in the Darwin home's culture of the truth seeking and discovery-loving biophilia.

Darwin mused about this questing in a letter to his son Horace. "I have been speculating [about] what makes a man a discoverer of undiscovered things. . . . As far as I can conjecture the art consists in habitually searching for the causes and meaning of everything which occurs." Science is all about searching for new truths, new ideas, and new observations. Attraction to the new, so basic in biophilia, also spurs the early adaptors among us, whose ranks include Emma. She embraced waltzing when society considered the dance risqué, and she was in the vanguard of medical practice when she added chloroform to her armamentarium for birthing her own children.

Emma and Charles delved into the truths in each other's hearts and minds, sharing and learning their personal polestars, as their letters show. Emma was grateful when Charles, early on, gave her what she called an "account of your own mind." Charles told Emma that the quest for knowledge, explanation, and truth was everything to him, writing: "The five years of my voyage . . . may be said to be the commencement of my real life, the whole of my pleasure was derived from what passed in my head." Emma's view of this? "I am sure it must be very disagreeable & painful to you to feel . . . cut off from the power of doing your work & I want you to cast out of your mind all anxiety about me on that point. . . . You must not think that I expect a holiday husband to be always making himself agreeable to me."

Emma and Charles lived by a tell-all pact. Emma wrote, "I will write down what has been in my head, knowing that my own dearest will indulge me." In another moment of trust, Emma wrote, "I do not wish for any answer to all this—it is a satisfaction to me to write it." Because she knew he was listening. Charles opened up in return, as Emma acknowledged, "I thank you from my heart for your openness with me. . . . My reason tells me that honest & conscientious doubts can not be a sin."

When Charles nursed Annie through her final weeks at a far-off health spa, he wrote expansively to Emma: "I . . . think it is best for you to know how every hour passes. It is a relief to me to tell you, for whilst writing to you, I can cry tranquilly." On Annie's death, Charles wrote, "We must be more and more to each other, my dear wife." Emma replied, "You must remember that you are my prime treasure." Charles felt "some sort of consolation . . . to weep bitterly together." They both looked to the good when in the grip of the bad. They both felt opposites coexisting. Their love grew from sharing this perspective, sharing of heart and mind and childcare, and sharing a commitment to the whole truth. Charles expected nothing less when he complained lightly to Emma: "Why did you not tell me how your old self was? Be sure and tell me exactly."

Darwin wrote to Henrietta of Emma as "our dear old mother who is . . . as good as twice refined gold." The love of Charles and Emma was so big as to be perceptible. A niece of Emma's wrote to her, "Your marriage made the strongest impression on me as a young girl and influenced me deeply in my ideal of married life. I felt from childhood your and Uncle Charles's exceptional happiness together." Another young relative, Snow Wedgwood, wrote to a friend that Emma's stance toward Charles was "almost the most remarkable I know in a wife, the union of absorbing devotion and perfect impartiality is so striking." In Emma—and in their marriage—love and truth went hand in hand.

CONSISTENCY UPSTAIRS, DOWNSTAIRS, AND IN PUBLIC

Charles and Emma expressed their biophilia outside the family as well. Both were abolitionists, for instance. Emma served her neighborhood with her skills, time, and money. She was, as noted earlier, raised to feel "responsible for the health, welfare and education of the poor in their area," shown when as a teenager she provided health care with her older sister. Marriage and motherhood did not stop Emma from doctoring in her new neighborhood or from continuing to teach reading and writing in the village. Emma also supported a lending library. She distributed vouchers for bread from the village bakery. She funded a cart to deliver water three times a week to a neighborhood one mile uphill from the nearest well, easing the work of girls and women.

Regarding the staff at Down House, which could number some two dozen, Emma's daughter reminisced, "She would take any trouble to help them or their relations, and in return there was nothing they would not do to please her. In an emergency . . . anyone would change their work; the cook would nurse an invalid; the butler would drive to the station, and anybody would go on an errand anywhere or be ready to help in looking after the poor people." Maybe this sounds too pretty to be true, but records support Henrietta. For example, when Emma's maid married the Darwin's main manservant and left Down House, Emma established a dressmaking shop for her. Emma then sent girls there to apprentice in needlework to enable them to earn money. And she paid the yearly school fees in the 1890s for at least one local girl who worked summers as her housemaid. Emma did not march with the suffragettes, but an employment program, the water cart, education for girls' empowerment, and the provision of childbirth services and health care are projects funded today by women's human rights groups.

Darwin eased life for the household staff when he renovated Down House, expanding and upgrading their quarters and facilities. He wrote in a letter, "It seemed so selfish making the house luxurious for ourselves and not comfortable for our servants." This is the Darwin who had long ago and in public condemned the treatment of slaves across South America. His travelogue, *The Voyage of the* Beagle, reveals his sense of common humanity with people of all cultures. For example, Darwin recounted a story from South America about some runaway slaves. He eulogized an old Indigenous woman who committed suicide as colonialists seized the group: "Sooner than be led into slavery, [she] dashed herself to pieces from the summit of the mountain. In a Roman matron this would have been called the noble love of freedom."

Locally, Charles and Emma crusaded against cruelty to animals; Emma shared Darwin's sense of shared humanity with other species. Together, Charles and Emma wrote a letter to their local newspaper, the *Bromley Record*, objecting to the traps used against vermin. And in the early 1860s, according to Henrietta, Emma pushed the national Society for the Prevention of Cruelty to Animals to offer a prize for a design for a humane alternative to steel traps. Emma later wrote about this trap in a public letter that was not ladylike but gruesome: "If we attempt to realize the pain felt by an animal when caught, we must fancy what it would be to have a limb crushed during a whole long night between the iron teeth of a trap, and with the agony increased by attempts to escape . . . [the] instruments of torture." Emma championed kindness to other organic beings.

Darwin sent the law after a man known to starve his sheep. Many people knew, Francis wrote, that "the two subjects which moved my Father perhaps more strongly than any others were cruelty to animals and slavery." As for the common practice among scientists of using conscious dogs in experimental research, Darwin wrote in the second edition of *The Descent of Man*, "Every one has heard of the dog suffering under vivisection, who licked the hand of the operator; this man, unless the operation was fully justified by an increase of our knowledge, or unless he had a heart of stone, must have felt remorse to the last hour of his life." The love of life shared by Charles and

Emma clearly entailed biophilia's feeling of unity with all life forms, respected as such by their peers or not.

THE SQUIRRELS WHO MISTOOK OUR MAN FOR A TREE

Darwin inspires hagiography, it is true. But biographers are not alone in succumbing to his charms. As the eminent novelist Henry James wrote after having lunch at Down House, "Darwin is the sweetest, simplest, gentlest old Englishman you ever saw. . . . He said nothing wonderful and was wonderful in no way but in not being so." Likewise, to other, less complex organic beings, Darwin was no imposing *Homo sapiens* but rather a chameleon, some ur being, humble and nothing wonderful. A man who shared snuff with a monkey imprisoned in a zoo. His son Francis recounted, "Sometimes when alone, he stood still or walked stealthily to observe birds or beasts. It was on one of these occasions that some young squirrels, mistaking him for a tree, ran up his back and legs, while their mother barked at them in an agony. . . . He always found birds' nests even up to the last years of his life, and we, as children, considered that he had a special genius in this direction." Day in and day out, because of his biophilia, Darwin resonated with every organic being around.

Emma knew her husband could be nothing but wonderful when he was with nature, writing in a letter: "Yesterday a wasp settled on Father's [Darwin's] face & put its proboscis into his eye to drink the moisture apparently. He got up very quietly from the sofa & stood looking at himself in the glass till the wasp moved. . . . A sting in the eyeball [would] have been horrid." For Emma, perhaps the fact that her husband was heretical was less important than that he was magical. As one biographer speculated, "It would take a special man to pry Emma away from home." If Darwin loved life, as she did, Emma saw that life loved him.

SELECTED BIBLIOGRAPHY

These books were key to my thought:

Bagemihl, Bruce. *Biological Exuberance: Animal Homosexuality and Natural Diversity*. New York: St. Martin's, 1999.

Barkow, Jerome H., Leda Cosmides, and John Tooby, eds. *The Adapted Mind: Evolutionary Psychology and the Generation of Culture*. New York: Oxford University Press, 1992.

Barrett, Paul, Paul Gautrey, Sandra Herbert, David Kohn, and Sydney Smith, eds. *Charles Darwin's Notebooks, 1836–1844*. Ithaca, NY: British Museum (Natural History) and Cornell University Press, 1987.

Barrow, John. *The Artful Universe*. Oxford: Clarendon, 1995.

Beer, Gillian. *Darwin's Plots: Evolutionary Narrative in Darwin, George Eliot, and Nineteenth-Century Fiction*. 2nd ed. Cambridge: Cambridge University Press, 2000.

Browne, Janet. *Charles Darwin: A Biography*. Vol. 1: *Voyaging*. New York: Knopf. 1995.

——. *Charles Darwin: A Biography*. Vol. 2: *The Power of Place*. New York: Knopf, 2002.

Cronin, Helena. *The Ant and the Peacock: Altruism and Sexual Selection from Darwin to Today*. Cambridge: Cambridge University Press, 1991.

Darwin, Charles. *The Descent of Man, and Selection in Relation to Sex*. London: John Murray, 1871. Princeton, NJ: Princeton University Press, 1981.

——. *The Expression of the Emotions in Man and Animals, with an Introduction, Afterword, and Commentaries by Paul Ekman*. 3rd ed. London: John Murray, 1872. Oxford: Oxford University Press, 1998.

——. *Journal of Researches Into the Geology and Natural History of the Various Countries Visited by H.M.S.* Beagle. London: Henry Colburn, 1839. New York: Hafner, 1952.

——. *On the Origin of Species by Means of Natural Selection, or The Preservation of Favored Races in the Struggle for Life*. London: John Murray, 1859. New York: Modern Library, 1998.

——. *The Power of Movement in Plants*. London: John Murray, 1880.

Darwin, Francis, ed. *The Life and Letters of Charles Darwin, Including an Autobiographical Chapter*. 2 vols. London: D. Appleton and Company, 1887. New York: Basic Books, 1959.

Dawkins, Richard. *The Selfish Gene*. 2nd ed. Oxford: Oxford University Press, 1989.

de Waal, Francis. *The Ape and the Sushi Master: Cultural Reflections of a Primatologist*. New York: Basic Books, 2001.

Dennett, Daniel. *Darwin's Dangerous Idea: Evolution and the Meanings of Life*. New York: Simon & Schuster, 1995.

Dewey, John. *Art as Experience*. New York: Minton, Balch & Company, 1934. New York: G. P. Putnam's Sons, 1979.

Dissanakyaki, Ellen. *Homo Aestheticus: Where Art Comes from and Why*. Seattle: University of Washington Press, 1995.

Fox Keller, Evelyn. *A Feeling for the Organism: The Life and Work of Barbara McClintock*. New York: Freeman, 1983.

Fromm, Erich. *The Anatomy of Human Destructiveness*. London: Jonathon Cape, 1973.

Gruber, Howard, and Paul Barrett. *Darwin on Man: A Psychological Study of Scientific Creativity, Together with Darwin's Early and Unpublished Notebooks, Transcribed and Annotated by Paul. H. Barrett*. New York: Dutton, 1974.

Houser, Nathan, and Christian Kloesel, eds. *The Essential Peirce: Selected Philosophical Writings*. Vol. 1: *1867–1893*. Bloomington: Indiana University Press, 1992.

Hyman, Stanley Edgar. *The Tangled Bank: Darwin, Marx, Frazer, and Freud as Imaginative Writers*. New York: Atheneum, 1962.

James, William. *The Principles of Psychology*. New York: Henry Holt and Co., 1890. Cambridge, MA: Harvard University Press, 1983.

Kuhn, Thomas. *The Structure of Scientific Revolutions*. 3rd ed. Chicago: University of Chicago Press, 1996.

Levine, George. *Darwin Loves You: Natural Selection and the Re-enchantment of the World*. Princeton, NJ: Princeton University Press. 2006.

Lumsden, Charles, and Edward O. Wilson. *Genes, Mind, and Culture: The Coevolutionary Process*. Cambridge, MA: Harvard University Press, 1981.

Maturana, Humberto, and Francisco Varela. *Autopoiesis and Cognition: The Realization of the Living*. Dordrecht: D. Reidel, n.d.

Miller, Geoffrey. *The Mating Mind: How Sexual Choice Shaped Human Nature*. New York: Doubleday, 2000.

Peckham, Morse. *Man's Rage for Chaos: Biology, Behavior, and the Arts*. New York: Schocken, 1973.

Phillips, Adam. *Darwin's Worms: On Life Stories and Death Stories*. New York: Basic Books, 2000.

Richards, Robert. *Darwin and the Emergence of Evolutionary Theories of Mind and Behavior*. Chicago: University of Chicago Press, 1987.

Santayana, George. *The Sense of Beauty: Being the Outline of Aesthetic Theory*. New York: Charles Scribner's Sons, 1896. New York: Dover, 1955.

Stevens, Wallace. *Wallace Stevens: Collected Poetry and Prose*. Edited by Frank Kermode and Joan Richardson. New York: Library of America, 1977.

Wilson, Edward. *Biophilia: The Diversity of Life*. New York: Library of America, 2021.

——. *Consilience: The Unity of Knowledge*. New York: Knopf, 1998.

INDEX